FINDING GOD IN THE WAVES

FINDING GOD IN THE WAVES

How I Lost My Faith and Found It Again Through Science

Mike McHargue

Foreword by Rob Bell

UNCORRECTED PROOF

CONVERGENT BOOKS
NEW YORK

Published in the United States by Convergent Books,
an imprint of the Crown Publishing Group,
a division of Penguin Random House LLC, New York.
crownpublishing.com

CONVERGENT BOOKS and its open book colophon
are registered trademarks of Penguin Random House LLC.

Library of Congress Cataloging-in-Publication Data
<~?~CIP data>

ISBN 978-1-101-90604-0
eBook ISBN 978-1-101-90605-7

Printed in the United States of America

Book design by Andrea Lau
Jacket design by
Jacket photograph by

10 9 8 7 6 5 4 3 2 1

First Edition

For Jenny and Mom, who held out a candle when I was in the dark.

For Madison and Macey, who already have questions on their own journeys with God.

And for my dearly departed granddad—the first to hear these words in our last afternoon together. After hearing them, he told me for the first time about his own doubts, paused for a moment, and then said, "Don't tell your grandmama."

The first gulp from the glass of natural sciences will turn you into an atheist, but at the bottom of the glass God is waiting for you.

—*Werner Heisenberg, theoretical physicist and one of the pioneers of quantum mechanics*

CONTENTS

FOREWORD

by Rob Bell

[TK—Please reserve 4 pages.]

[TK—Please reserve 4 pages.]

FOREWORD

[TK—Please reserve 4 pages.]

[TK—Please reserve 4 pages.]

INTRODUCTION

She looks at me with haunted eyes.

We're in a small church held in a hotel conference room in Texas, and I've been talking about God and the brain for almost an hour. I just answered a question about how Christians should talk to atheists, and as soon as I finished talking, she spoke up.

"So, what I'm hearing you say is that it's OK to be an atheist and come to church and pray. Because I've been sitting in church for, maybe, four months now, and I've been an atheist my entire adult life."

She tells me about studying anthropology in college, how she came to find all this God stuff preposterous. And yet, she likes to pray. She says it gives her comfort. But then she asks a question that makes my heart ache.

"Can I sit here in church like an impostor?" she asks. "Am I an impostor?"

I take a moment to collect myself. Her honesty and vulnerability are too familiar: I, too, have sat in a roomful of

Christians and admitted I don't believe in Christ—or in any God at all.

I tell her a story about a man walking along the shore of a lake. On his way, he runs into two fishermen. They're busy working, but he holds their attention long enough to tell them he'll show them how to bring in people instead of fish if they come with him. The two fishermen drop their nets and follow the man.

I tell her one of those fishermen was Simon (who is also called Peter) and that he is one of the founders of the Church (with a big *C*). When Peter dropped his net and followed the man—Jesus—he didn't know anything about the Messiah being a sacrificial lamb or about crucifixion or resurrection. He just heard the man's story and believed it enough to follow him.

The Gospels are a collection of stories about Peter and the other 11 disciples constantly doubting, believing the wrong thing, or entirely missing the point of what Jesus was saying.

So, do I think it's OK not to know what you believe and still be a part of the Church?

Heck, yeah. In fact, I think that's exactly what following Jesus is about.

We live in interesting times, don't we?

We've got atheists in churches and Christians who would never dream of stepping into a house of worship. Four in ten Americans believe that God created Earth with His own hands fewer than 10,000 years ago. Three in ten believe that the universe is billions of years old and that life developed via evolution without any intervention from any god. The rest of us generally believe some combination of those two extremes.

We're also living in a period of tremendous social upheaval. The secular nations of Europe have, for the most part, bid farewell to organized religion, although belief in things spiritual and in the afterlife remains common. The United States is among the most deeply religious nations in the developed world, and yet even here religion is in historic decline. The fastest growing American "religion" is "none."

More congregations are shrinking than growing, and more churches are closing than starting up. But that doesn't mean atheism has become more popular. Most Americans say they would never cast their vote for an atheist president, and commonly held scientific insights involving climate change and the Big Bang theory remain controversial with the American public. Meanwhile, however, atheism has become one of the largest and most organized "religious" movements on the Internet.

For me, this isn't simply a sociological trend or mere matter of religious history. It's personal. Throughout my childhood and young adulthood, as a member of a conservative Southern Baptist church, I loved and followed Jesus, until my faith unraveled when my parents' marriage fell apart. The collapse of their 30-year union sent me to the Scriptures, where I thought I'd find God's answers to the crisis. But instead of answers, I found contradictions and a God so brutal it frightened me. In time, I became an atheist in private while in public continuing to serve as a church leader, deacon, and Sunday-school teacher. Afraid of being found out, I led a double life: showing up at church on Sunday mornings, then going home afterward to plan a post-religious future for the world with other atheists on the Internet.

I was the world's least interesting double agent. But my mission was killing me.

Years later, God moved miraculously in my life, and I came

back to the fold of the faithful. But it wasn't by studying sacred scriptures or works of theology. It was through science—studying neuroscience and cosmology and discovering a God who was as mysterious as quantum physics and as intimately near as the neurons of my own brain. That journey is the story of this book.

I know what it's like to be a Christian. I have been saved and baptized, and I have followed Jesus day by day for years on end.

I know what it's like to feel that God is standing next to you, and I know what it's like to hear His voice.

I also know what it's like to doubt.

I know how it feels to have questions pile up until their cumulative weight crushes everything you thought you knew about the world and its Maker.

I know what it's like to be an atheist.

I know what it's like to trust only in what you can support with empirical evidence.

I know what it's like to know right from wrong without the aid of divine laws, instead relying on careful examination of how human actions can violate others' consent and produce suffering.

These streams of faith and doubt, religion and science, collide in our culture, creating rapids and whirlpools that rob people of their sense of meaning and purpose. Our different beliefs about God tear families apart, fuel culture wars, and even drive some people to suicide. This volatile mix of faith, doubt, secularism, spirituality, and atheism have driven our American melting pot past the boiling point. But I've also learned that it's possible to reclaim two seemingly staunch adversaries—science and faith—as partners, to open myself to the movement of God without rejecting scientific insights about our world.

You can know God intimately while acknowledging the mystery, even the absurdity, of such a notion. You can experience the proven neurological benefits of prayer even as you contemplate how science shows prayer's limitations. You can be part of the global body of people who follow God without turning off your brain or believing things that go against your conscience. You can read the Bible without having to brush off its ancient portrayal of science or its all-too-frequent brutality.

And you can meet a risen Son of God named Jesus while wondering how such a thing could ever be true.

PART I

LOST AND FOUND

CHAPTER 1

Dinosaurs in Sunday School

I was a fat kid.

I had no discernible athletic talent. I wore my hair in a bowl cut and had an odd appreciation for Hawaiian shirts. The shelves in my bedroom were full of computer magazines, spare parts from robots, and toys from science fiction movies arranged in scenes of battle.

Here's how deep my "nerdery" ran: When I was eight, I took apart a VCR and reassembled its parts in a lunch box. I put the lunch box in a backpack and then ran some cabling from the backpack to a roughly cylindrical mechanical assemblage that I had scavenged (OK, *stolen*) from my grandparents' farm in rural North Florida. The end result was a homemade proton pack, which allowed me to start an unlicensed Ghostbusters franchise in my neighborhood. I convinced my friends to build packs of their own, and we would roam the streets of our neighborhood at night, catching ghosts. I had no idea how prophetic this would be, my fixation with a movie in which humans dominated the supernatural with science and technology. But that's a story for later.

For now, all you need to understand is that, in the 1980s, a passionate love for science, an overactive imagination, and a chubby physique were not exactly the recipe for popularity. I was a round peg (a *very* round one) for a too-small square hole, and this made my grade-school days a living hell.

My elementary school in Tallahassee, Florida, was like a John Hughes "Brat Pack" film gone horribly wrong. Ever since I could remember, an unofficial but strict hierarchy had dominated our social world. Everyone knew who our leaders were: a small collection of boys who were the funniest, the fastest runners, and the first picked when we played team sports. I both idolized and feared them.

The rest of the social pecking order was indecipherable to me. But I knew I was at the very bottom, the nerdiest of the nerds. Time spent in my company was damaging to anyone's reputation—and, in fairness to the other children, it's not as if I hadn't earned my social standing. People usually picture nerds as introverted, maybe even antisocial. Certainly many are, but I think some are like me: extroverts of such intensity that it makes others uncomfortable.

I once told my classmates that I was a werewolf—a fact about which I was absolutely convinced.

Then there's the fact that I cried at the drop of a hat, something other grade-school boys take in with the excitement of a shark smelling blood.

At recess, tag was the worst. I ran like someone wading

through molasses, and my classmates knew that once I was "it," there was no way for me to transfer that dishonor to anyone else. When the alpha kids discovered this, they began running backward and chanting, "water tank, water tank," making fun of the way my belly made waves when I ran. I was easy prey—a fat, ginger gazelle in "husky" jeans.

By second grade, every recess had come to represent a choice: I could try to play with other kids and be bullied, or I could seek solitude and make it through without tears or having to call for teacher intervention.

So I chose solitude. Each day when the recess bell rang, I would make a beeline for the woods at the edge of our playground, where I would pass the time inventing stories to tell myself. This strategy wasn't 100 percent effective. Occasionally a teacher would fetch me from my hiding place because I'd ventured too far afield; other times, a bored bully would actively hunt me down. But more often than not, I was out of sight and out of mind, and they left me alone.

Alone.

And lonely.

I became a Christian when I was seven.

My family comes from the largest denomination of the conservative Evangelical movement: the Southern Baptist Convention. Southern Baptists believe that people become reconciled to God when they believe that Jesus is the Son of God, who died for the sins of all of humankind, and when they state that belief in a prayer.

That prayer is called the Sinner's Prayer, a name that says a lot about what it's like to grow up in that wing of the church. Southern Baptists believe that all people are born sinners. Humans are in love with pleasure, power, and prestige, and our natural inclination is to follow our sinful natures into all sorts of trouble. This isn't too far-fetched an idea. I carry 40 pounds of evidence of this tendency around my midsection.

But most Southern Baptists take it further, believing that people are completely hopeless without God, and that anyone who isn't saved through faith in Jesus goes to hell—an actual, physical place of eternal, fiery torment and suffering. This concept can do a number on the imagination of a seven-year-old kid, which was the age my friends and I were when we heard it. Many children express interest in salvation right around the time they're old enough to grasp this concept of eternal torment. Some of my friends remember having nightmares in which their "unsaved" friends roasted in fiery pits while they looked down from heaven's paradise.

I'm thankful to report that this wasn't my experience. I was fortunate to grow up in a congregation that focused on the hope of salvation—a message that was more carrot than stick. People at my family's church talked about having Jesus in your heart and the Holy Spirit in your life. God was someone who helped you make the right decisions, understand the Bible, and find peace no matter what was happening around you. That sounded wonderful to my small ears.

One night after coming home from church, I interrogated my parents about salvation. Even as a kid, I was never the kind of person who accepts information without scrutiny, and I wanted to see if I could find or poke any holes in this salvation concept. I don't remember this conversation, but my mother tells me it was remarkably businesslike. I wanted to know how, exactly, the process worked. What words did I say to be saved? What did God do, exactly, when I said those words? How would I know that God was doing His part? After nearly an hour, my curiosity was sated, and I went to get ready for bed.

I usually fall asleep quickly, but I couldn't that night. I felt a sense of urgency, an energy pulsing through my bones. I knew I needed to ask Jesus into my heart, so I grabbed my mom and told her it was time—that I was ready to know Him. Mom asked a few questions and then led me in that Sinner's Prayer as we knelt beside my bed—an altar covered with Snoopy sheets.

A few weeks later, the congregation baptized me, my teeth chattering in a baptismal pool with a broken heater. My feet didn't reach the bottom of the tank, so I dog-paddled to the pastor and stood on his boot. Moments later, after I was dunked under that frosty surface in the name of the Father and of the Son and of the Holy Spirit, the preacher said I was a new creation, and I felt it. I was inspired, and I couldn't wait to share the good news.

So I didn't. The next day, I went to school and told every one of my classmates that I loved them. Every one of them. Every last boy and every girl—all of them so equally horrified by my pronouncements of love that I found myself in the principal's office.

My faithful walk with Christ wasn't helping my social standing at school. Maybe I took that "ye are . . . a peculiar people" thing too literally.

But my faith did help me in other ways. When I felt lonely hiding in the woods to escape bullies, I would talk to Jesus. I talked to him about feeling fat, slow, and stupid. Sometimes I would ask him why, if he truly loved me, he had made me the way he did. Other times, I asked Jesus to make me able to hit a home run or run a mile without stopping, and I would imagine the admiration and accolades that would come from the other kids when such a miracle happened. I didn't think it was too much to ask. Jesus was God, after all, and God had parted a sea for His people. All I was asking for was one lousy home run.

I never got that home run, but at least Jesus was a good listener. He never made fun of me, either.

Our talks weren't all lament and pleading. We had a lot of fun, too. We'd talk about how the world worked and all the things in nature that amazed me. I didn't have any friends at school, but that was OK. My best friend lived in my heart.

These days, people often tell me I'm smart. Every time I hear it, I'm amazed, because no one made that assumption during my first few years of school.

I had a hard time learning to write and spell. Around the time my classmates were forming legible letterforms, my scrawl still looked like preschool graffiti. And even though I loved to read, my spelling was atrocious—bad enough, in fact, that I was put in a special class for a few hours each week. It was a strange class, one that housed both the smartest kids and the kids who had trouble learning.

My parents kept having to come to school to talk with my teachers about my unrealized potential. The teachers would tell them that I needed to work harder and apply myself. A couple of them said I was smart but lazy. I believed them. I hated myself for being so lazy.

At the end of each school year, my grades were usually good enough for me to advance to the next level but bad enough to initiate a serious talk about holding me back. Each year this talk got a little more serious. I'd probably still be in the third grade had it not been for a miracle that saved my academic career.

My school got computers. And computers changed everything.

In those days, computers were expensive and unproven, and they weren't kept in every classroom. They had their own special domain—a small room where the Apple IIs, with their green-on-black screens and giant floppy disks, were kept in two rows.

Computers and I became fast friends. You could press a key, and a letter would appear on the screen as if by magic. I felt none of the frustration I usually experienced when forming handwritten letters. Before long, I was crafting words and sentences with ease. This didn't free me from the tyranny of penmanship, but it at least helped my teachers see that I wasn't a hopeless case.

Before long, I had taught myself the basics of programming by modifying the educational video games the teachers gave us to play. I would figure out how to name fish after myself in a game called Odell Lake or get extra money in The Oregon Trail. One day, I created a program that would write my name on the screen over and over again. Everything about the machines made sense to me, because their abstract, procedural way of thinking resembled my own.

Computers are the reason people think I'm smart. They're confusing and mysterious to a lot of people, but I was hardwired to thrive in cyberspace—more suited to that world than the one everyone else enjoys.

Seven was a big year for me. I met Jesus, who saved my soul, and computers, which saved my life.

Programming computers, taking apart VCRs, and building ad-hoc proton packs all came from the same root: I've always wanted to know how stuff works. I loved taking things apart and studying their components. I loved to discover what each piece did and how it contributed to the function of the whole.

Humans are innately curious; our minds are driven to build models of the world that help us not only find food and shelter, but also predict the future. We build those models by learning from our experiences and by hearing the experiences of others. Our survival is linked to this ability, and it creates in us a craving for certainty just as powerful as an ant's craving for sugar.

In school, I learned that this craving could be satisfied through the discipline of science and its methodical approach to probing reality. I loved science then, and I love it now. It laid bare all mysteries, and my grasp of it was the one thing that made me feel superior to my classmates, cousins, and even my parents.

You may have been able to throw a baseball, but I could calculate its arc. You could set fire to a pile of leaves, but I knew that combustion released the energy those leaves had captured from the sun. You might have gotten your first kiss years before me, but I knew why you got that fuzzy feeling when you did.

Science textbooks offered me arcane knowledge my peers ignored—and so did the Bible. It was a potent drug, this feeling of greater knowledge, but it also had an unfortunate side effect. Sometimes, it turned me into the thing I hated most: a bully.

Though I had few friends at school, three boys my age lived on my street: Todd, Jon, and Ketan. Like me, all three were more interested in *Star Wars* than football and were more likely to spend a day exploring the woods than playing a pickup game of baseball. These guys were my refuge, the only place I could experience belonging with my peers.

Jon and I spent a lot of time together. He's incredibly smart—in fact, he's one of the smartest people I've ever known. Geniuses like Jon are the reason I chuckle whenever some kind person tells me that I'm the smartest person she knows. I had this odd ability to remember a lot of things, but Jon could drive all those memorized facts into fresh ideas about how the world works. I could read books that were advanced for my age, but Jon was always many levels ahead of me. I don't mean *grade* level. I mean that if I was reading high-school-appropriate materials in grade school, Jon would be consuming stuff meant for college students. I'd get excited about Plato's Allegory of the Cave, and then Jon would tell me what Nietzsche had said about it. I would pretend to know who Nietzsche was because I hated to feel dumb.

Jon made me feel dumb all the time. Even worse, Jon could draw really well, and I could barely manage a stick figure.

One day, when Jon and I must have been nine or ten, we were talking about God. This was a big moment, because Jon's family didn't go to church, and he didn't talk about God very much, but now, while we stood in my driveway, he was being open and sharing his beliefs. Jon said he thought God was like a filing clerk. His desk was full of cards, and those cards composed our reality. This divine Clerk had cards for physics, biology, historical events, future events, people, and everything everywhere. God spent his time filling out these cards and keeping them organized.

I laughed at him.

I couldn't help it. Jon's idea about God was so ridiculous and wrong. After all, I was an expert on God—I studied him every week at church, and I even knew him personally. I'd been saved and baptized. Here was Jon, so much smarter than I, and he didn't understand God at all. I have to admit, the thought made me feel pretty proud of myself.

After I stopped laughing, I started to tell Jon about the one true God of the Bible and His Son, Jesus. But Jon didn't want to listen. His eyes were watering, and his face was red, too. I couldn't tell if he wanted to cry or yell at me. I think a lot of us do what I did in that moment. We have experiences with God that are beautiful and moving, but over time, they just become things that make us feel superior to other people.

I don't remember what he said next. I only remember that he went home, and I was left standing in my driveway, wondering where I'd gone wrong. Jon doesn't believe in God at all these days, and I can't shake the feeling that I might have had something to do with that.

Just before high school, two things mercifully conspired to transform me from an ugly duckling into—well, not a swan, but at least a slightly cuter, hipper duck.

First, thanks to a flood of hormones, I grew taller with such speed that in order to feed its metabolic furnace, my body had to burn off its prodigious belly fat. Second, one of my friends from middle school told me I should learn to play the bass guitar. He was learning to play the guitar, and he thought that if we started a band, people would like us.

Girls might even go out with us. Maybe.

It was the romantic equivalent of a get-rich-quick scheme, but it worked.

At first our band was terrible, but we kept at it, and soon enough, a few hundred people would show up every time we played. This was the early '90s, the era of grunge. The airwaves were full of "Smells Like Teen Spirit" and "Jeremy," and pop culture celebrated the quirky, the odd, and the downright weird. All of a sudden, my accidentally oversize shorts were cool, and my shaggy hair could be stretched into a hip (in those days) ponytail.

For the first time in my life, I had friends. Lots of friends. Even football players and cheerleaders were talking to me. For a kid who'd spent his early years hiding from everyone else at recess, this sudden rush of attention and acclaim was a powerful drug. I craved more of that validation, and I got it.

Still, a part of me always thought that the whole thing was a big prank. At any moment, someone would throw dog food on me or make fun of me in front of everyone, revealing that my newfound acceptance was just an elaborate setup. I even had nightmares in which all the guys in my band pointed at me and laughed.

What a bummer.

In many ways, our pain and our way of coping with it define who we are. These experiences shape us and mold us, for better and for worse. They can compel us to help others or drive us to numb the pain in whatever way we can.

For me, romantic intimacy became that numbing agent.

When I finally started dating, I would cling to my girlfriend the way a shipwrecked sailor might cling to a buoy in the middle of the sea. I was needy to the point of utter dependence. Having someone who wanted to be near me somehow made me feel whole, secure, and complete.

It was against this backdrop that I discovered sex. Baptists believe that sex outside of marriage is a sin, and I had never had much trouble following any of God's other commandments. I read my Bible, prayed often, and when I went to parties, I didn't drink. But sex was different. Even the personal connection I felt with God couldn't wipe out my loneliness in the way that physical intimacy could.

My life became a sine wave of transgression, forgiveness, and redemption. I was crushed by guilt every time I "went over the line." Each time I would swear it was the last, a resolve that was like a sand castle trying to stand against a rising tide.

I moved out of my parents' home soon after high school graduation. I celebrated this freedom by entering a series of increasingly unhealthy and destructive relationships. I couldn't stand to be alone, whether it was two in the afternoon or two in the morning, so I was drawn to any woman who'd offer me the magic elixir of her company. Suffice it to say, girls from church didn't want to hang out at 3:30 in the morning—and they certainly didn't want to be alone with me overnight.

By this time, I was regularly playing gigs in bars—shows that went into the wee hours of the morning. The women I met in those venues were happy to do what girls from church wouldn't.

I started hitting the snooze button on Sunday mornings, and then I stopped setting the alarm at all.

Some of the leaders in my church began to notice my absence. One staff member even started warning people my age that I was a bad influence—an accusation to which I responded with righteous indignation.

I decided I was done going to church. Jesus, I knew, wasn't confined to any one building. I could follow Jesus on my own, without being judged by a bunch of hypocrites.

There was only one problem with this resolution. There was one woman I wanted more than any other, and she was a good Baptist girl.

From the first time I laid eyes on Jennifer Carol Frye, I was smitten.

At the time, I was a lowly high school senior, and she was a college girl. Her eyes shone like spotlights when she smiled, and she smiled a lot. She was funny and snarky; plus, she looked great in a tight T-shirt. When my final high school romance fizzled, I set my sights on Jenny Frye.

I was stunned when my plan to woo and win her didn't work out. That may sound conceited, but I assure you, I mean it in the way an engineer is stunned when a bridge she designed falls down. You see, my advance from nerdery to popularity wasn't by accident. I'd learned to play guitar, sure, but I'd also taught myself to be funny by watching comedians. Even more, I had carefully observed the mannerisms, conversational cadences, and body language of the most popular, confident people around, and I'd welded those same attributes onto my own personality.

In time, I found I could get people to like me if I wanted them to—especially women.

With Jenny, I deployed every weapon in my arsenal—well-timed jokes, casual touches, and room-capturing antics. She'd respond with a smile or wink, but she refused to go out with me. She was not interested in buying whatever I was selling. I'd walk up to say hello, give her a hug, and then lunge her into a deep dip, à la Fred Astaire. She'd laugh politely but nothing else. Jenny Frye was looking for a man, and in a lot of ways, I was still that clueless, needy, ginger gazelle.

I didn't give up. I found every excuse I could to hang out with Jenny: going bowling with her friends, even though I hated bowling, or showing up at her house unannounced to take her to dinner. I did this even when I was dating other people. Finally, it began to work. We became best friends, and not without good reason. Studies have shown that proximity is one of the primary drivers of human attraction. We really do love the one we're with.

All our time together added up, and eventually Jenny gave in. We went out on a proper date, and then another and another. But she wouldn't kiss me, much less do anything more than that. Jenny was less concerned with theology, philosophy, and the cerebral parts of faith than I was, but she was way better at living what she believed. I could be as bad an influence on her as I wanted to be; she wasn't affected.

She was furious every time I missed church. The morning after my late-night shows, Jenny would sit through the worship service and stew in her anger while I slept at home. The first thing I'd experience on a Sunday morning was an angry call or, worse, a furious knocking on the front door.

She was right to be angry. Every Saturday, I'd tell her I would be at church on Sunday morning. Every. Single. Saturday. I gave Jenny a lot of lip service about her being the most important person in my life, but what I really meant was that she was the person my emotions had the strongest response toward. If she was actually important to me, I would have backed that up by keeping my word—or at least by being honest.

Finally, Jenny couldn't take it anymore. She told me if I wanted to be with her, I needed to be at church. If I wouldn't come back to our church, she told me we could find a new one. But we had to go somewhere, and we had to go together.

I took Jenny's words and wishes to heart. I stopped telling her she mattered enormously to me and instead started acting as if she did. I started going to church again, too.

Because Jenny wouldn't let me live a different life during the week than I lived at church on Sunday, this was the start of a golden age of sorts for me. For the first time, I started to really pay attention to sermons and Sunday-school lessons. I became more honest with people and less likely to lie to impress them. I began studying the Bible on my own—not to impress Jenny, but because I wanted to learn from it. I even joined a Christian band, trading late-night bar gigs for Sunday-morning worship and church camps.

Jenny and I got married on a record-breakingly cold December day. Before long, I was teaching Sunday school at our church and playing worship music at both the students' service and the Sunday-morning service for adults. My life was full of older, wiser people whose counsel I could seek, and I did so whenever

I felt the need. I was ordained as a deacon in the church when I was 25—an honor I took seriously. One year later, Jenny and I became parents to our first daughter, Madison, and her little sister, Macey, followed two years later. Although I was already a demonstrative person, becoming a father caused my heart to swell. Every day, I thanked God for my life and the people I got to spend it with.

People are down on Evangelicalism these days, but even my earliest years of life showed me that Evangelical churches are great at doing a whole lot of important things. They provide community, comfort, and stability. When an active member of an Evangelical church dies, the family of the departed receives immense support during their grieving. Dealing with the influx of casseroles and baked hams delivered to the homes of the bereaved can become a logistical issue, and their grass is mowed as if by elves.

What I'm saying is that it's easy to stand on the outside and dismiss Evangelicals as crazy Fundamentalists, but this misses most of what the movement really is (or, at least, is supposed to be). I'm not an Evangelical anymore, but it was Evangelicals who showed me how to be a loving husband and father. Evangelicals taught me how to be a good employee and how to live my life with integrity. And Evangelicals were there for me when my life fell apart.

—

CHAPTER 2

Binge Reading the Bible

I grew up in a household that was straight out of a 1960s sitcom. Dad made the money, mom kept the house running, and my younger sister, Melissa, and I fought over issues of great importance, such as whose turn it was to take a bath first.

Our stable, loving home was a safe haven during the isolation and bullying of my early school years. My parents loved us unconditionally and abundantly. They were both affectionate toward my sister and me, and they showed me what it looked like to have a strong marriage. I could always count on my parents not only for wise advice, but also for an example I could eventually follow in life and in marriage.

So, of course, I was wary when my dad called and said we needed to have a family meeting. What was a family meeting? We'd never had one before. I thought about what could prompt a gathering with such a serious label. Cancer? Could one of my folks be sick? Or did Dad lose his job? Whatever it was, it couldn't be good.

I remember walking into my parents' house that night. The tiny foyer seemed to stretch beyond its normal length. As I turned into the living room, my sister and Mom sat on the couch, an empty seat between them, while Dad sat alone in a recliner. I sat in the remaining empty recliner and surveyed the room.

Mom avoided my gaze, and Dad seemed nervous. My sister and I conveyed our mutual ignorance with a glance. I don't remember, but I may have tried to make a joke to lighten the mood. Then Dad started a speech.

He told us he was leaving Mom. He said they'd still be friends, but he'd fallen in love with another woman, and while he'd fought that as long as he could, "the heart wants what the heart wants." Honestly, the specifics of what he said are hard to recall, because I started to go into shock. I couldn't believe what I was hearing.

You have to understand that I always idolized my father.

Dad is a "manly man" of the highest order. A couple of years ago, Dad was impaled—I mean literally impaled—by a fallen tree. He was driving a tractor to clear some brush on our family farm outside the tiny town of Greenville, Florida, when a tree somehow got kicked up between the tractor's wheels and their guards, impaling him with a branch the size of a grown man's arm.

Dad has quick reflexes. Before the tree could pass all the way through his torso, he put the tractor into reverse and used it to pull the branch out of his body. Our farm is in the middle of nowhere and has very little cell-phone coverage, so Dad had to drive the tractor back to the barn, where he bound his wounds with towels and duct tape and then drove himself to the tiny hospital in town. His injuries were too grievous to treat at the local hospital, so they cleaned his wounds, loaded him into an ambulance, and rushed him to the larger hospital in Tallahassee, straight into surgery.

That's the kind of man Dad is. He's made of steel and leather. He can build buildings and fix mechanical things, and he has a particular affinity for sweat, grease, and dirt. Dad played football, basketball, and baseball. Dad shoots guns. I, on the other hand, regularly clean my computer keyboard with alcohol and Q-tips because I shudder at the thought of my fingers getting dirty.

In many ways, my father and I couldn't be more different. We didn't understand each other while I was growing up. Dad tried hard to introduce me to sports, but I screamed in horror any time a football was thrown my way. On the few occasions when he grudgingly agreed to play video games with me, we discovered that he was absolutely hopeless with a joystick. We stood on either side of a fundamental divide—Dad was a jock, and I was a nerd—but we loved each other, and I looked up to Dad more than anyone in the world.

Now the man who had taught me how to be a husband, father, and man of God was telling me he'd decided to end his marriage.

At first, I couldn't understand why Dad would want a divorce. My parents had been married almost 30 years. They seemed happy. They were always flirting and putting their hands on each other when I was young. Mom and Dad were a unified front when it came to parenting. They always backed each other up. They seemed perfect for each other.

When my initial state of shock passed, the gears and cogs in my skull started to spin at incredible velocity, performing spiritual and ethical calculations. Up to that point, I'd always submitted to my father's God-given authority, but in that moment, it was clear that Dad had failed his charge. He'd broken an essential commandment. The Bible is clear: Divorce is not part of God's plan. So I spoke up.

"Dad, are you a Christian?" I asked.

He said that he was.

"Well, if you're a Christian, your life is not your own. You have been bought at a price, and you belong to God. Do you understand that?"

Tentative and hesitant, he said that he did. I began to pontificate, full of righteous authority. I asked Mom whether she wanted to end the marriage or try to work it out. Mom said she wanted to work it out, that she still loved Dad, so I told Dad that he couldn't get a divorce because he didn't meet the biblical requirements for one. As long as Mom was willing to forgive and accept him, he had to work toward resolution and reconciliation.

Looking back now, it seems so crazy. I was a man in his twenties, married less than a decade, telling two seasoned adults with nearly 30 years of marriage under their belts how to live their lives. I was so certain that I had it all figured out. I loved my wife "as Christ loves the church." I taught my children but didn't "provoke them to anger." I worked hard at my job "as working for the Lord, not for human masters." I followed God, and he'd given me a better life than anyone deserves to have.

I remember leaving my parents' house thinking, "God can fix this." All we had to do was obey Him. My dad had obviously been charmed by the devil, but I knew that the power of Satan was no match for the power of Jesus.

"The Joy of the Lord is my strength . . ."

I had always thought *joy* meant the minimization of suffering. I hated to feel bad. I've always had a naturally sanguine disposition, but this is at least in part because I can go to pretty unhealthy lengths to keep the boat from rocking.

When faced with a problem, I took a simple approach. I'd think of the three people in my life who knew the most about the type of problem I was dealing with, and I'd ask them what to do. Relationships, spiritual issues, and the like all led me to the same people: the ministers at my church.

The pastor was out of the office the day I went for counsel, so I sat down with Brother Rick, the associate pastor, and Cory Pitts, the student pastor. Rick was about my dad's age, with the quiet assurance and gentle disposition of someone who has wrestled with demons and won. Cory was just a few years older than I. He's the kind of guy who buys flowers for his wife "just because," who can take his son fishing in the morning before work and his daughter to a dance at night, all without getting too busy or too tired.

I sat down to tell Rick and Cory what had happened, but instead of laying it out in clear, concise statements, I cried and stuttered. The safety of that office had uncorked the grief I'd been unconsciously holding down. They had to give me tissues, because I couldn't get my emotions under control. I just kept sobbing.

People don't go to church as much as they used to do. Every year, fewer and fewer people show up on Sunday mornings. I totally get it. Churches can be really crummy, judgmental places. Churches have hurt a lot of people, as in really, deeply injured. Plus, have you ever gone to brunch on a Sunday morning? It sure beats getting up early and putting on fussy clothes to sit in uncomfortable pews for an hour and a half.

But at their best, churches provide incredible support for people in their times of greatest need.

Church is a place where you are guaranteed to get a hug when you are hurting, where someone will tell you he's sorry for your loss. It's a place to hold hands, sing songs, and celebrate the gift of life and living. It's a place to be broken and to heal and for others to tell you they are broken, too.

I wonder what would happen if churches focused on those things, instead of websites, growth strategies, and building campaigns.

Imagine a church that was safe instead of slick.

Maybe we wouldn't have empty pews. Maybe it would be standing room only. Again.

It took a while, but finally I was able to stop crying and tell my pastors about my grand plan to save my parents' marriage. I laid out all the key points in Scripture, all the things I would say to stop my dad from rebelling against God. I told them I was going to be his accountability partner, and I outlined my plan to take him through the Scriptures into a proper understanding of God.

It felt good to be in control again.

At least it did until I noticed the sad look in Rick's eyes.

"You know, Mike," he said, "sometimes after someone has an affair, the family makes them apologize so much, and for so long, that the person who cheated doesn't feel like a part of the family anymore. They're the ones who have broken a covenant, and they often can't forgive themselves. If the family can't forgive them, either, reconciliation is impossible."

Then Cory started talking about how lifeguards swim carefully near drowning people, because too often the drowning person will grab his would-be rescuer and use her as leverage to try to climb out of the water. It's the survival instinct. When people become panicked and exhausted, they simply try to survive. Period.

Then Cory added, "Mike, your parents have been in the water a long time. You just jumped in. You may be a strong swimmer, but I'd be careful trying to carry them. This divorce will affect you, too. You will have your own feelings to deal with. If you don't take time to cope with this situation, your parents' breakup could cause you to doubt yourself as a husband and a father."

Then he handed me the book *To Own a Dragon*—writer and public speaker Donald Miller's account of growing up without a father and all the emotions that came along with that. Cory told me that he knew my dad was a great father but that I might start to feel abandoned by him. He thought reading Miller's book might help me deal with some of those issues (especially since another Miller book, *Blue Like Jazz: Non-Religious Thoughts on Christian Spirituality*, had played a big part in my faith).

Then my pastors gave me the name and number of a Christian counselor who specialized in divorce. I asked whether I should pass the information along to Dad.

"Mike, that number is for you."

I was wholly bewildered as I left the church that day. Rick and Cory seemed to think I was the one in danger, not my dad. They also seemed to doubt my plan to save my parents' marriage. That didn't sit well with me at all. Still, I decided to heed their suggestions: I would read the books and see the counselor. If nothing else, it would keep me strong enough to do the work of saving Mom and Dad.

When I read *To Own a Dragon,* I cried more than I ever had in my life; it was often hard even to see the pages. Cory had lent me the book, but I had to buy him a new copy when I was finished reading it; my tears had stained too many of the pages.

Around this time, I decided I should look for additional counsel, so I went to the smartest human being I have ever known: Stratton Glaze.

Stratton, a former business associate of mine, is as tall as a southern pine and just as thin. There is much more to Stratton than his tallness, but his tallness is so overwhelming that it has to be mentioned. Stratton is an introvert; he only speaks when he has something to say. But when he does, he tosses out blinding insight in the manner that other people talk about the weather.

I was Baptist and Stratton was Methodist, but in a flourish of ecumenical détente, for a time Stratton and I shared an office while working to start a social-media marketing operation for the ad agency I worked for. We did all the usual things guys do when they share an office, such as discussing philosophy or talking about how technology would influence civilization in the future.

Stratton and I saw the Bible differently. I believed the Bible was inspired and inerrant, which is Baptist-ese for saying that God wrote the Bible, and, therefore, it is perfect. In this way of thinking, the Bible is accurate in whatever it talks about, including science and history. If the Bible says the whole world was flooded thanks to 40 days of rain, then that's exactly what happened. Although the Bible may not speak to every topic someone could face, the Bible did offer the principles for making any decision in a way that would please God.

That's not how Stratton saw the Bible. He said the important thing about the Bible was its stories and the way they revealed the nature of God and how He relates to people. Stratton told me that the Bible's narrative was the important part and that trying to describe God as a logical system was a flawed approach. God, after all, is beyond any system of human thought.

That last part sounded good, but the rest of it really challenged me. After all, I had once made fun of a friend for saying, "God is bigger than the Bible."

When I told Stratton about my parents' breaking up, I was pretty upset. As I said, Stratton is quite reserved, so I remember being surprised at the way his eyes watered as I spoke.

He listened carefully without interrupting me—then he said the most surprising thing I'd ever heard. Stratton told me that God was going to use my parents' situation to change me and grow me. He was sure that God was working in my life.

Don't read that wrong. Stratton wasn't saying that God caused my dad to have an affair. He was saying that God was going to use it anyway.

As with so many theological ideas, this one might appear to represent strange logic, but it can be incredibly useful for finding peace and meaning when traveling life's darker paths.

If God is all-powerful and good, why does evil exist in the world? People have given this topic a lot of thought for a long time. They've thought about it for so long that they've even given it a technical name: theodicy, or a defense of God's omnipotence in view of the existence of evil.

People "solve" the problem of evil in different ways. Skeptics solve it by eliminating God. A lot of the really awful things in our world make more sense if the universe is inanimate, life a fortuitous series of coin-flips in cosmic probability. Cancer, sickness, and natural disasters are all just the product of physics being physics, natural laws unfolding in a way that increases entropy over time. The universe doesn't love us or hate us: It's infinitely indifferent. Even the more active forms of evil—rape, murder, and genocide—are just leftovers from the latent brutality out of which civilization arose. Some human brains didn't get the memo that we no longer live in small bands of viciously competitive tribes.

Believers, on the other hand, tend to see evil as the result of humanity's free will. God gave us agency and the ability to act on our thoughts. For some reason, we have a tendency to choose things that hurt others if they help us. Christians call that a "sin nature" and say it extends all the way back to the Garden of Eden and fall of man, as described in the book of Genesis. In this way of viewing the world, God is there, and he loves us, but he's letting us learn and grow by making our own choices.

Skeptics say that's all rubbish, that if parents stood by in the name of free will and watched as one of their children strangled another of their children, those parents would go to prison. This is a really good point and one for which I've never heard a good answer.

Despite its logical failures, the idea that God watches us lovingly from a distance, that good actions will ultimately be rewarded, and even that good will one day triumph over evil gives comfort to billions of people in terrible circumstances.

It even compels people to take action for good causes against impossible odds.

It's an illogical mess. But when you apply it to life, it somehow works.

Dad and I talked on the phone almost every day. He'd tell me that he couldn't understand how the love he felt for the other woman could be bad. He'd say that he desired to reconcile with Mom but that he couldn't escape the gravitational pull he felt toward his new lover.

I had answers for all his questions, but it troubled me that my scriptural insights didn't miraculously deliver him back into a right relationship with my mom and with God. What if I had misunderstood some key fact? What if my advice to my dad was tainted by my flawed, human perspective? What if the verse that would win my dad back to God was waiting for me, right there, in some untouched corner of Leviticus?

I'm a nerd, and nerds view all problems as an opportunity to conduct research—so I decided to get familiar with the entire Bible, to go deeper than the guided tours I had taken via studies in the past. I read the Bible from cover to cover in three months. That may sound shocking, but it's easier than you'd think. Just Google "Bible in a year plan," pick a plan, and read four days worth of readings every day. At my reading speed, this meant about 30 minutes a day at a relaxed pace.

When I got to the end, I started again, back at the beginning. I'd never read the Bible all the way through before, but in that fraught year of the divorce, I did it four times.

But reading the whole Bible didn't help. In fact, it made things worse. All my life, I had known that the Bible conflicted with history and modern science, but I had notions about how to reconcile those disagreements. When I took a look at the Hebrew Scriptures (a.k.a. the Old Testament), though, it made me question everything I knew about God and the Bible.

The problems began, appropriately enough, in Genesis 1.

I had enough working knowledge of science to understand why scientists believe that our universe is very old—roughly 13.8 billion years. I also understood why scientists believe it all started with a Big Bang.

Neither of those things troubled me. It's no stretch to imagine that a Big Bang might have been what happened when God said, "Let there be light." And the term for "days" in Genesis 1—the Hebrew word *yom*—isn't always translated as *day*. *Yom* can be translated as a year or even an epoch, depending on the context. Who could say how long each of these "days" was? I had no problem believing each could be billions of years.

But this time around, I was reading more closely, and I noticed other scientific conflicts. The first problem I had was with the idea of God creating a "firmament" and calling it heaven. When I looked up what a firmament was, I found out that the ancients believed it to be either a vast dome or a series of interlocking plates that separated Earth from the heavens. The stars, they posited, were holes in the firmament where the light of heaven shone through, and these same holes were the source of rain and snow.

But we've been into space. There is no firmament in the heavens.

I tried to imagine cosmology from the perspective of someone living 3,500 years ago, and I decided that several concepts in astrophysics could be related to a firmament, such as the edge of the visible universe or maybe even the Higgs field.

But I also noticed that Genesis 1 says trees were made before the stars, a notion so absurd that I had to reread the passage several times to make sure I wasn't missing something. Here's why:

Stars live for hundreds of millions to billions of years. Because of their intense gravity, bigger stars burn faster, while smaller ones (such as our Sun) can be stable for billions of years. Stars don't actually "burn" at all—they're much too hot for combustion. Instead, stars are so massive that the force of their gravity causes atoms to smash together and fuse, releasing massive amounts of energy in the process. Over time, the fusion that happens in a star's core makes heavier and heavier elements, until eventually it starts producing iron. No star is big enough to fuse iron at the atomic level, so once iron shows up, the star can't produce fusion anymore, and it dies.

Stars have spectacular deaths. The big ones explode, and when they do, they cast heavy elements out into space. Those elements become asteroids, planets, and dust. This dust becomes rocks, algae, trees, animals, and people.

Genesis says we were formed from dust, but cosmology tells us that you don't get dust—unless you have stars first. Without dust, you don't have the material to make trees or humans. There were no trees in our universe before there were stars.

Of course, God could have made a universe in which trees

lived before stars. But a universe that started with Earth and trees would look different now. The elements on Earth aren't anywhere near as old as the universe itself. Our planet is a spritely 4.5 billion years old in a universe that has existed for close to 14 billion years.

I guess God could have made Earth and trees first and then hidden the fact by messing with atomic isotopes on Earth and crafting a universe that looks much older than it is. But why would God do that?

Why would God make His Holy Word and the night sky tell two different stories?

Less than a day into my Bible-reading plan, I was already discouraged. I don't know how many times I've read Genesis, but its contrast to cosmology had never seemed so stark to me before. I stopped to say a prayer, asking God for guidance as I continued reading His Word in search of peace and wisdom.

There's a whole Bible, after all—1,189 chapters in 66 books—and here I was, splitting hairs over the opening paragraphs. I decided to trust God and just keep reading. In fact, I had a hunch that some later passage would help me reconcile what I knew from science with what I read in Scripture. If not, God is God, and science must be wrong.

Except Genesis 2 was worse than Genesis 1.

It told the same story but in a different sequence. This was a logical impossibility.

God is all-knowing.

God wrote the Bible.

The Bible is perfect, inerrant, and infallible. It is free of contradictions.

Genesis 1 says that God made plants, then animals, then people.

Genesis 2 says that God made Adam, then Eve, then plants and animals.

Impossible! How had I missed this? I'd read the book of Genesis countless times, never picking up on this issue of chronology. My stomach turned cold, and my face felt flushed as I pondered an idea that troubled me deeply: It's one thing for the Bible to contradict science, but it's something else entirely for the Bible to contradict itself.

God can't contradict God. Either God made people on the sixth day, or he didn't. The first two chapters of Genesis seemed to offer different answers from each other, and that wasn't possible. Every spiritual leader I knew had told me that the Bible was without error or contradiction. My parents' marriage depended on this core truth, as did my faith.

Of course, this wasn't the first time something in the Bible had troubled me. The Bible is a complex document, and people have always struggled to interpret parts of it. An entire branch of theology exists to deal with this tension: apologetics. Scholars who can read the original languages and have knowledge about the context in which Scripture was written help laypeople (such as me) understand some of the Bible's more challenging ideas and passages.

I looked up the opinions of several apologists, and their answers to the Genesis 1 and 2 problem was consistent: Genesis 1 is about how the world was made, while Genesis 2 is about the sixth day of creation and how God made the Garden of Eden.

Viewed in this light, there is no contradiction. Either God made plants and animals for the garden specifically, or Genesis 2 is referencing the work done in Genesis 1.

Sure. Except I still had one problem: Why didn't the Bible itself say that? Why would God Almighty leave such a major point to the apologists? I said a prayer, and a line of Scripture came to me.

His ways are not my ways. His thoughts are not my thoughts . . .

After all, it's right there, in the Bible. Who was I to challenge my Maker for His artistic choices? I took some time to pray and apologize to God. I asked Him to help me with my pride.

After praying, I decided to start a spreadsheet on which I could keep track of verses that seemed to hold contradictions. It was just too arduous a process to stop and read apologetics every time I hit a passage that didn't line up in my mind.

I kept reading. I read the story in which God drowns all the life on Earth except for the lucky few who got put on an ark. I hadn't really read Genesis deeply since becoming a father, but now I imagined my daughters drowning because of my wickedness. The Bible says that mankind had become wicked, but I couldn't picture how infants or toddlers could be wicked enough to deserve that kind of death.

I read about God hardening the heart of an Egyptian pharaoh, then killing all the first-born children in Egypt because of that hard heart. I read about God ordering a newly free nation of Israel to purge the Promised Land of its existing inhabitants—with explicit instructions to kill every man, woman, child, and even the livestock.

It was right there, in black and white. God ordered His people to commit genocide. The same God who "so loved the world that he sent His only begotten Son" ordered soldiers to kill infants and burn innocent animals so that His chosen people could have their own land.

This was not the loving God I knew.

This God was terrifying and brutal.

How had I missed all this before?

The only thing worse was the answer proffered by apologetics: that God had a right to do with His creation as He pleased.

To me, that God seemed like a sociopath.

From there, my prayer life became fervent and intense.

Every day, I asked God for help. I told Him that I felt lost and confused, and the more I read the Bible—His Word—the more confused I became. The Old Testament was a torrent of violence, incest, murder, war, rape, and senseless death. And my spreadsheet of contradictions was growing faster than I could look up answers.

I expected to find relief when I got to the New Testament Gospels, those beloved stories about Jesus' life on Earth. The words and actions of Christ were encouraging, at least when they weren't confusing and inscrutable. I mean, why is Jesus cursing a fig tree or saying that men should castrate themselves for the Kingdom of God if they can handle it? But the Gospels had the same problem as the earlier books: They didn't line up with one other.

Matthew, Mark, Luke, and John offered me different facts, figures, and orders of events.

And Paul. What was his problem with women? Why did he think marriage was to be avoided?

The whole Bible was opening itself in a new light, but not a good one.

There was no way this book was inerrant.

Forget inerrant—this book seemed downright immoral.

It was just as Pastor Cory had said. In my deep dive into the Scriptures, I forgot all about trying to help my dad get back to God.

Now I was the one who was drowning.

There was no way I could voice these kinds of objections in church. Baptists call themselves "people of the book," and questioning the Bible's authority is tantamount to treason. Instead, I started to pester my colleague Stratton Glaze with questions at the office.

He told me I was reading the Bible all wrong. He said that what I was going through was one of the problems with systematic theology. I was trying to see God as an equation to be solved instead of a living being who partners with us in His creation. I was reading His Word as a physics or history textbook instead of an epic tale about God, man, and the relationship between the two.

Stratton challenged me to read a book called *Velvet Elvis: Repainting the Christian Faith*. It was written by Rob Bell, a pastor who had started a big church in Michigan. Stratton really liked Bell. He seemed creative, and although some of his teachings had concerned me in the past, I wasn't sure I could survive another round of questioning God's Word on my own. So one night, when I was in San Francisco for a business conference, I sat down in a chair facing my hotel window and started reading *Velvet Elvis*. My eyes were wet before I finished the first chapter.

Velvet Elvis presented faith from a new perspective. Bell

described the work of Jesus as being ongoing, rather than as a set of revelations and customs that were set in stone thousands of years ago. He openly discussed the doubts I felt and the logical incongruities I encountered, but not as a way of tearing down Christianity. Instead, he clearly saw those inconsistencies as evidence that God is continually working to produce new work and insights in people.

It was like a breath of fresh air. I realized that my approach to God and his Word was static and confining, like leaving a toy in its box to display on a shelf instead of enjoying it as it was meant to be. I felt closer to God and less afraid of what I was reading in the Bible. The Bible was inspired, yes, but it was also the work of humans. And humans had opinions, fuzzy recollections, and agendas. If God's Word was filtered through humans' words, it makes sense that the flaws of humans would be found in the text.

Looking at the Bible in this way meant that I didn't have to see contradictions as being evidence against God. Instead, they were reflections of a God confident enough to share Himself through the words of humans. This was a God of redemption, hope, and mercy—co-creating the world alongside His people.

When I came home from the trip, my Sunday-school lessons got better. I was more passionate than ever about what the Bible could teach us. Cultural context and critical reading showed me lessons I'd never before seen in its pages. I knew I'd found a more powerful way of understanding God.

I also stopped fretting over my parents' marriage. It wasn't my job to fix them. It wasn't my job to point out the sins of others. My only job was to love my mom and dad with all my heart and to offer them healing however I could.

It felt freeing to stop believing I had to defend God and instead embrace the task of being God's follower. I had no doubt that I had finally figured out God, the Bible, and how to live the best life.

"Pride goes before a fall" is a cliché for good reasons.

CHAPTER 3

A Foolish Bet

Christians can be really obnoxious. But you probably already knew that.

I'm not sure what can be done about it. We are a people who believe that Jesus died to save the whole world, and some of the most famous words attributed to Him command His followers to go tell the whole world about it. Our zeal to see you "saved" can be really tiring if you're a Buddhist, Jew, Hindu, or secular American.

Muslims understand our zeal, at least. Their faith demands the same drive to conversion.

My sensitivity to the feelings of others and my desire to obey Jesus were always at odds with this imperative, and the better I got at getting people to like me, the more this tension increased. I was always afraid that forcing my faith on someone would lead to rejection. On the other hand, what if my fear of speaking up caused someone to end up in hell?

I compromised by living the best life I could, while taking

every opportunity I could to advertise that I was a church person. My theory went like this: If my life was wonderful enough, and if I was loving and helpful enough, surely people would connect my peaceful centeredness with my love of God.

Here's the crazy thing: It actually worked.

People asked me to pray with them all the time. They came to me with their problems and questions about God. I took this all as validation that I was doing what God wanted me to do.

I prided myself on having friends outside the church and being a good friend to them, too. Even when I was sure I had the right answers about life, I was careful to be respectful of other people's beliefs. This was partially pragmatic: You can't lead someone to Jesus who won't talk to you.

But it was in this spirit of cooperation that my faith died.

My friend Tom is an atheist and secular humanist.

It's not just that he doesn't believe in God—Tom believes that belief in God is actually harmful. I can't think of a better adjective for Tom's atheism than *evangelical.* He basically goes around preaching the gospel according to Darwin. For the 12 years we've known each other, Tom and I have kept up an ongoing, spirited discussion about faith, morality, and spirituality.

Oddly enough, I've never met Tom in person. That's because we "met" online in the late '90s, a time when you had to be a computer expert to get on the Internet. We were regulars on the same gaming service, a lobby where players sought opponents to join them in their favorite video games.

When Tom found out about my move toward Progressive Christianity (which emphasizes compassion, social justice, tolerance of human diversity, and responsible stewardship of Earth's environment, often through political activism) and my newly open views on the Scriptures, he was fascinated. For the first time, I was starting to come out on top in our discussions about the existence of God. It was much harder for Tom to dismantle a Bible that was the "Word of God through words of humans" than a Bible that was the perfect work of God via divine Dictaphone.

Then, one day, I was explaining morality in the context of Shalom as a Jewish religious principle when Tom interrupted me by asking, "Have you heard of Richard Dawkins?" It was an odd question in the context of our conversation, but I didn't miss a beat. "Sure, I know Dawkins. I read *The Selfish Gene* in high school and it convinced me evolution was true." I felt very cool and enlightened. Nonbelievers were always surprised when I said I accepted the theory of evolution.

But Tom continued. "Have you read his latest book?" he asked.

Tom's question let me know where this conversation was headed. I'd seen Dawkins on on *The Daily Show* talking about his latest book, *The God Delusion.* Dawkins's book was big news because it argued that religion wasn't just wrong—it was actually dangerous. I remember scoffing at the biologist's contention that a supernatural Creator surely did not exist, because I knew that all true science reveals God. I remember thinking that Dawkins should stick to biology and leave faith to the faithful. I told all this to Tom, but he wasn't convinced. This time *he* didn't miss a beat.

"Listen, the way you approach faith these days seems really reasonable," he wrote. "I'll make you a deal: I'll read a book of your choosing about the way you understand God if you read *The God Delusion.* We can compare notes after we've read each other's books."

I didn't respond immediately. On one hand, I felt better than ever about my relationship with God. On the other, I'd only recently stabilized after the shock of my parents' divorce and the spiritual vertigo that my scriptural binge sessions had brought on. What if *The God Delusion* started that process again? I didn't want to wrestle through another round of crippling doubt.

But this was about as interested in God as Tom had ever been. I was sure that if I could get him to read *Velvet Elvis,* his resistance toward faith would dissolve. This was an opportunity for me to close a sale for Jesus. Plus, I already knew all about science. I understood the Big Bang theory and Darwin's theory of evolution, and I'd integrated them into my faith as being part of God's handiwork. What more could Dawkins have to share? How could this particular scientist shake a science nerd's faith?

I accepted Tom's challenge and bought *The God Delusion* for my Kindle. I was really confident. I trusted that God's truth couldn't be broken by any book; if nothing else, this would be a way to grow closer to God. I expected to come out of this deal a stronger Christian.

Instead, reading the book was like stepping into the ring with Mike Tyson. I would come out alive but bloodied, bruised, and broken.

The God Delusion is a slow burn. Most books come out swinging with their best material in the introduction, but the introduction and first chapter of *The God Delusion* didn't say anything I hadn't heard before. I expected the rest of the book to be an easy-to-dismiss series of straw men and half-baked arguments.

But then Dawkins talked about prayer. He referenced a series of scientific studies that had failed to show any positive effect of prayer on people recovering in hospitals. The study seemed to be well conceived, executed, and documented. Based on this platform of data, Dawkins tore down the idea that God acts on the prayers of His followers.

Prayer was the foundation of all my beliefs about God. Even though I could no longer accept the Bible as a literal, inerrant work, prayer still sustained me. The Bible was still powerful when studied prayerfully, using its stories to reflect on my life, its Psalms as prayers to God in good and bad times. It was in prayer that I felt closest to God, in prayer that I heard Him speak to me. Prayer had introduced me to Jesus, and prayer had helped me grow to know Him better, back when I was hiding from bullies in the woods.

Most of all, I had seen prayer work. I remember when the doctors told us that my Dad's mom (I called her Mema) was at the end of her days, that this time the cancer would not give up. We all prayed for her, and in a few months her tumors were gone.

Another time, God compelled me to pray with a friend. I didn't know why God wanted this, but I felt pulled by some invisible magnet, and I knew I wouldn't have peace until I did so. So I called my friend and asked if we could pray together, "right now." I drove to his office, walked in and closed the door, and started to pray words that seemed to come from somewhere else.

As I prayed, tears started to fall from my friend's eyes. His head was bowed, so instead of rolling down his cheeks, his tears puddled at his feet. I didn't know precisely why I was there to pray, but the act proved profoundly resonant that day.

I had seen the power of prayer with my own eyes. How could science tear that down?

It turns out that I'm not the first Christian who's argued for God's existence by talking about answered prayer. In fact, atheists have an experiment designed to challenge anyone who thinks of God in this way.

They ask you to pray to a jug of milk.

Imagine you wanted a promotion at work. So you prayed to a milk jug for two weeks and then applied for the promotion. If the promotion came through, would you say it happened because the milk jug answered your prayer? Most people would not.

Most Christians say that God answers prayer in three ways: *yes*, *no*, or *wait*. If God says yes, you get whatever you were praying for. If God says no, then you don't. If God says "wait," then you keep praying for your desired outcome, knowing that God's timing is different than your own.

But the problem is, that covers every possible outcome. Things either happen now, later, or not at all. There's no other possibility. How can you be confident that prayer works if there's, literally, no scenario that could prove it to be false?

It seems obvious, but I'd never thought of it that way before. The more I thought about it, the more I realized that God had never answered a prayer of mine. My grandmother's cancer did go away, but it's not unheard of for cancer patients go into remission without therapy. I wasn't necessarily witnessing the hand of God; I was witnessing probability and ascribing God's hand to it. This realization made me feel silly and superstitious.

So I stopped thinking of prayer as a hotline to God or as

some magical means of changing my circumstances. Instead, I resolved that prayer was simply a way of connecting to God—that mysterious, powerful energy that caused everything. I started to see prayer as a way to change who I was and to drive me to act according to God's will.

That seemed wise, even noble. But it also made me suspicious of God.

If God is real, why doesn't He answer prayers more directly? Why is He always so distant and mysterious?

If we are His beloved creations, couldn't He make it more obvious that He's out there and that He sent Jesus to save us?

I read the rest of *The God Delusion* and then moved on to dozens of works from other skeptics. Each one introduced me to new arguments that challenged God's existence and cemented the ultimate conclusion: There is no evidence that God exists.

The works of Christian writers all emphasized faith. Faith is the belief in things unseen.

But these writers can't really mean simply unseen.

Gravity is unseen, but we can measure it. Air is unseen. So is love.

All those things are part of the physical world. We can test them, prod them, and document them. We can find their origins and clearly measure their interactions.

Not so with God.

You certainly can say that if God is a supernatural being, all-powerful and beyond our universe, such a God could act as He pleased. But any time He interacted with the physical world, He would leave evidence.

If God really burned a water-soaked wooden altar at the request of Elijah, a scientist on the scene could have documented the rise in temperature and the combustion of the wood. If God really appeared as a pillar of cloud by day and a pillar of fire by night, the water vapor in the cloud and the searing heat of the flame could be measured, documented, and verified.

But none of the miracles claimed by modern Christians left any such signature. Modern miracles either are urban myths that can't be traced to an original source, or they are easily dismissed frauds. Faith healers might claim to banish back pain, but researchers have found that this "miracle" often involves convincing the petitioner that one of her legs is shorter than the other by sliding one shoe slightly off the heel to simulate the longer leg, only to slide the shoe back in place so the petitioner sees herself "healed." This visible "healing" and the effects of adrenaline primed by the "healing" experience mask the pain.

It wouldn't be hard to authenticate a genuine faith healing. All you'd need is a qualified doctor to evaluate a patient's injury or condition before and after the healing takes place. Yet has this simple test ever been done to demonstrate this particular power of God? Not that I've heard of.

Belief in things unseen is one thing; belief in things without evidence is another. And this latter kind of belief has a problem: How do you know what to believe? If God reveals Himself to us, but we have no empirical way of verifying that revelation, how do we know if that revelation is correct? Catholics, Protestants, Jews, and Muslims all claim that God has revealed Himself to them in different ways. Most people believe in the revelation that is most established in their family or community.

But imagine for a moment that you didn't know anything about God.

How would you choose among all the various world religions?

Each would tell you to have faith and that God would reveal himself to you. It would tell you that if you trust God, He will reveal the truth to you.

Now, imagine if I said that God revealed to me that people can fly if they imagine they're being lifted by angels. Would you step off a roof based on my revelation? I'd bet you wouldn't. You'd ask me to prove it. Asserting that people can fly is a serious claim, radically different from all your prior experiences. You'd want proof before believing something that far-fetched.

And yet here I was, believing that God had guided people through the desert for 40 years with pillars of cloud and fire. I believed that God lived in a tent and then a temple. I believed that the son of a virgin rose from the dead. Those things seem even more incredible than flying people.

The more I read, the more lost I felt.

Have you ever noticed that Christians and atheists don't talk *to* one another as much as they talk *at* one another? I used to think this was because the two groups disagree on one point of central importance: Christians are certain there is a God, while atheists are certain there is not. But this adversarial stance comes from a deeper well than simple opposition. My turmoil at reading the works of skeptics showed me that believers and unbelievers view the world in fundamentally different ways.

If you want to see what I mean, watch a debate between a prominent skeptic such as Richard Dawkins and a Christian apologist or clergyperson. Most of the claims the skeptic makes will be based on measurable, physical evidence. If Richard Dawkins tells you he believes that the Sun is five billion years old, he'll point you to a pile of data and measurements that support that belief.

He'd tell you that if you measure the Sun's mass, size, and luminosity and reconcile that with what the Sun is made of and our understanding of nuclear fusion, you'd get an accurate date value. That value is similar to the value we get when we date asteroids and Earth itself via different means. The fact that different means of measurement produce similar results is why we have such confidence in our Sun's age.

Science doesn't ask you for faith. It's happy to show you its homework.

Believers, Christians included, accept evidence as a way to learn about the world. But they also believe you can learn about the world through revelation—that God Himself can show you truth. Christians say that the universe itself reveals God, as does our need to know Him. Christians cite the Bible as proof of God's existence and power, an authority derived from its many prophesies and the "witnesses" of the miracles contained within it.

Skeptics say, "But how can you prove it?"

That's the rub. Christians believe that God is self-evident, and skeptics see no earthly evidence of God at all. For the few hundred years that atheists have been around in significant numbers, human society has yet to find a way to bridge this gap.

At least, not without crossing it.

As my faith unraveled, I kept praying. I was convinced that God would rescue me, even as I grew increasingly desperate to reconcile belief in Him with the insights and information I was learning from skeptics. I sought truth. I strove to know God better. God is the author of truth, so I was certain I would catch God if I chased after the truth. God was good, and God loved me. We'd get through this.

My prayer times became a series of nearly panicked requests for guidance, wisdom, and understanding. I asked for humility and for grace. I begged for some sign that God was there and that He was listening.

My theology was changing, sure, but God was taking me somewhere. I knew it.

The human brain is the most intricate and mysterious arrangement of matter in the whole known universe—at least according to human brains. Your brain has about 86 billion neurons and trillions of supporters called glial cells. (We know this number because a team of researchers used chemicals to dissolve brain tissue without damaging the nuclei of neural cells.) Our 86 billion nerve cells create connections to one another via dendrites, which are like tiny organic wires that allow neurons to send and receive electrical signals.

These signals, along with corresponding chemical messages, are the stuff all your thoughts and feelings are made of. Every song you've sung, every dream you've had, and every conversation you've lost yourself in originated as a burst of electrical activity in the trillions of connections among billions of neurons in your brain.

Human brains are massive. So big, in fact, that our babies have to be born at an earlier stage of development than do the offspring of other mammals; otherwise, passage through the birth canal would be impossible. That's one of the reasons human babies are so helpless compared to those of other mammals—our bundles of joy aren't fully baked yet. Even with early parturition, human childbirth is very risky to the mother. Take away modern medicine and we'd have a higher risk of maternal death during childbirth than most mammals, although we fare better than some species of monkey and hyenas.

Biologically speaking, human brains are also really "expensive." Your brain can't move, but it consumes up to 20 percent of your nutrients and 25 percent of your oxygen. This is why humans' brains are hidden inside thick, dense skull bones. Your body is designed with the purpose of protecting and supporting its most important organ.

Ancient people believed that our heart and bowels were the seat of our thoughts and emotions, but today we understand that our thoughts and feelings originate and transpire in our brains. There is nothing that is more "you" than your brain. No other organ has the capacity to shape how you perceive the world or how your inner experience unfolds.

So while it's your eyes that sample photons that bounced off your wife's face, transforming the resulting image into electrical signals that zoom along your optic nerve, it's your brain that decodes that raw data and creates not only a recognition that the object across the table is your wife, but also the feelings of love and affection you have for her.

People tend to view their minds and spirits as being distinct from their brains, but research doesn't support that idea. If you break your arm, you don't change much—a cast and some inconvenience usually are the extent of recovery. A heart attack is terrifying, but those who survive are shaped more by their fear and recovery than they are by the actual injury.

But if you injure your brain, you change dramatically.

Just ask Henry Molaison. He fell from his bicycle in 1939, when he was seven.

Something happened when his head hit the ground. In the weeks and years following the injury, he was plagued by severe seizures. These proved so dangerous that doctors intervened surgically when he was 27, removing most of his hippocampus and a number of other structures in his brain in a desperate bid to stop the seizures.

The hippocampus is part of the limbic system. It looks like a seahorse (or, rather, seahorses—you have two), and the word *hippocampus* is Greek for "horse sea monster." Scientists believe that the hippocampus plays a key role in memory formation, and Molaison is one of the reasons they do.

Although surgery successfully eliminated his seizures, Molaison no longer could form new memories. Later in life, he was shocked anytime he looked in a mirror—he expected to see a young man in his twenties looking back. He was stuck in the past, unable to build a new self-image based on memories.

Imagine walking up, looking in the bathroom mirror, and seeing yourself 30 years older than you expect. How would you feel? Of course, Molaison never stayed upset for very long. Within minutes, he couldn't remember having seen his reflection.

Then there's Phineas Gage, a 19th-century railroad worker whose left frontal cortex was mostly destroyed when an explosion on his jobsite drove a metal rod into his skull. Afterward, Gage's friends said he was "no longer Gage," and people described him as animalistic and unable to control his own impulses. Following his accident, his speech was profane and vulgar in all social contexts. He was impatient, prone to chase whatever desire he felt in the moment. Today we understand why this happened: The prefrontal cortex plays vital roles in consciousness, decision-making, and impulse control, and the damage to Mr. Gage's prefrontal cortex had basically taken it out of the equation.

Most tragic is Charles Whitman, who in 1966 murdered his wife and mother before climbing a tower at the University of Texas, from which he shot and killed 16 people and wounded another 32.

Whitman was a former U.S. Marine, described by his friends as being polite, friendly, and intelligent. But, prior to his outward display of violence, something in Whitman's inner landscape changed enough that he sought mental-health counseling. Just before his murderous rampage, Whitman left a suicide note describing a recent surge in "many unusual and irrational thoughts."

In the note, Whitman also requested that an autopsy be performed on his corpse. He thought there might be a biological basis for his violent impulses and frequent headaches. His request was granted, and doctors found a pecan-size tumor next to his amygdala—the part of the brain responsible for fear, anger, and aggression. The demons plaguing his mind turned out to be a mass of tissue pressing against the most volatile part of his brain.

Henry Molaison, Phineas Gage, and Charles Whitman all are members of a long procession of unfortunate people who have shown us how important our brains are. Everything that makes you who you are happens in your brain. Your thoughts, feelings, beliefs, dreams, hopes, and fears all are held ethereally in a pattern of organic "wires" and chemical bonds.

You are your brain, and your brain is you. And that means spirituality and religion are rooted in the brain in the same way that thoughts and feelings are.

Scientists know there is a part of your brain responsible for anger and another part responsible for affection. Your brain has a "spot" for language and a "spot" for vision, so some neuroscientists naturally wondered if the brain also had a "God spot," a part of the brain that's responsible for religious experiences. Despite several published findings, no scientific consensus has been built around the idea—there doesn't seem to be any one part of our brain responsible for God.

Still, whatever you know about God and whatever spiritual experiences you've had are held in your brain. More sophisticated brain-imaging technology has shown that people's beliefs about God aren't anything like a "spot" but instead arise from a complex network in our brains. The more someone thinks about God, prays, or has other spiritual experiences, the more developed this network becomes.

This is why belief in God is so robust in the minds of many Christians. God is not something we believe in as much as something we feel and experience—and this is why the faithful and the skeptical find it so difficult to understand one another. In the brains of atheists, *God* is a noun, a noun no more real than *tooth fairy* or *unicorn.* But believers have a rich neurological network that encapsulates God through feelings and experiences that are difficult to articulate with mere language.

Because I had prayed every day for most of my life, the roots of God in my brain were deep—my neurological "faith network" developed with years of practice. The problem was, that network wasn't keeping up with the new model of reality that my brain was building.

Your brain is a machine of sorts that builds a model of the world by throwing away most of its sensory input. I'm about to explain this in more detail, but it has a troubling takeaway: The way you see the world is a lie. Your senses and your brain work together to produce a false narrative about reality to your consciousness—a false narrative without which you would be unable to function. You take in a truly staggering amount of sensory information, far more than you could ever process, and your brain manages this overload by tossing out anything unrelated to your survival and creating a summarized view of reality that it can read on the fly.

If you don't believe me, just look at your TV or smartphone with a magnifying glass. The images and videos on these screens are nothing more than a bunch of tiny flickering lights. But your brain lies to you, saying what you actually see is a sheriff walking out of a hospital in an episode of *The Walking Dead.*

Your brain is doing you a favor. Long before television, our ancestors had to look at a field of waving grass and deduce whether a lion was hiding among those stalks. Human brains filter our all that extra movement to give the conscious mind an awareness of threats and rewards.

Then comes perhaps the most complex narrative that our brain and senses sell us: personal consciousness.

Have you ever wondered where you come from? I don't mean what town you live in or where you were born. Nor do I mean who your parents are. I mean, what's the difference between you and a toaster or you and an earthworm? Why are you something that can wonder why you exist, a function that a toaster or an earthworm could never begin to perform? To explain this, let's look to an unlikely example: the U.S government.

Our federal government is one of the largest organizations in the world, with more than 4.5 million employees. This scale allows the organization to do a lot of important things. It can turn empty land into roads, send robots to other planets, and maintain the largest military in human history, all because it is made up of millions of roles, departments, and moving parts.

Now think about the president. The president is the chief executive of the federal government—but his control is not absolute. The president obviously has to contend with Congress and the public, but even within the executive branch, his control and knowledge are limited.

Just ask the president to build an F-22 for you. Or an aircraft carrier. Heck, most presidents probably couldn't tell you where to find an extra pencil in the White House.

Thanks to the scale of the government, no one person knows how to do everything the government has to do. The knowledge and responsibilities of making the government run are distributed across millions of people who are experts in their individual roles.

The president's role is to make high-level decisions based on the information provided to him by his advisors, staff, and cabinet. That information is a summary of reports that come from all the departments in our government, and even those reports are a tiny fraction of all the information moving around inside those departments. It happens this way because government could not function if the president had to analyze every decision in full before signing off.

But the president's powers are even more limited. If someone fires a gun on the White House lawn, the Secret Service won't ask for the president's approval or opinion on what to do. Agents will immediately surround the president, grab him or her by the jacket, and move the country's leader to a secure location. They'll do whatever it takes to keep the most powerful person in the world safe. The president's veto power is only returned once the Secret Service is certain he's safe.

Now imagine if the entire U.S. government was housed in one massive building. Instead of the White House, the Capitol building, the Pentagon, all those NASA sites, and thousands of embassies and military bases all over the world, imagine that every single employee worked in one building that could hold four and a half million workers. (To give you some sense of scale, the Empire State Building holds about 21,000.)

That's your brain. That's you.

The president of your brain is the prefrontal cortex, a patch of tissue right behind your forehead. Most people's primary consciousness happens in the left pre-frontal cortex, others' the right, but either way, one side takes charge. This is the part of your brain with the highest concentration of neurons, the part that gives you the ability to perform rational analysis.

Your prefrontal cortex has the soul of an accountant. It is cold and rational. In its eyes, life is a spreadsheet, and every decision requires a careful examination of risk and benefit. This is where your conscious awareness emanates from—a patch of tissue right behind your forehead, mostly on the left side.

Like the president, your prefrontal cortex doesn't have absolute control of your thoughts, senses, and emotions. For one, it's slower than the more ancient parts of the brain where feelings originate. When you sense that you're in danger, your limbic system takes over. This is because fear and anger are cognitive shortcuts. You don't take the time to ponder the ethical ramifications of your actions when you're in danger—you act quickly and do what you have to do to survive.

When you recall those actions later, you'll remember them as if you were consciously in control the whole time. But that's a lie—in that moment of crisis, your limbic system will have bypassed the prefrontal cortex. And such unconscious actions aren't restricted to emergency situations. You don't have to think consciously in order to breathe or even to catch a ball tossed unexpectedly in your direction. Your prefrontal cortex doesn't analyze the speed and trajectory of the ball and make a ruling on where your hand should move. Your visual cortex and motor cortex just do what they do and tell the president about it later.

Like the U.S. president, your prefrontal cortex only gets a summary of information coming into the brain, no more than is necessary for making a decision. You'd never be able to get anything done if you had to concentrate on making your heart beat, while simultaneously planning every specific muscle action require to write the letter *A*.

Because your consciousness emerges from so many different parts of the brain, it is not always a cohesive whole. Like the U.S. government, it emerges from millions of smaller entities going about their business. And like the government, all those millions of parts in your brain don't always agree with one another.

The president may want to protect the environment, but the Senate majority may want to create policies that advance business interests. Likewise, your prefrontal cortex may make a rational decision that it's time to lose weight, but the hypothalamus will tell you it's best to eat that hamburger right now, while it knows the calories are available. It takes a long time to change your mind, because your brain isn't a computer. It takes time to rewire the networks of neurons in your head.

And these networks of neurons are important in other ways, as well. When you think of your mom, different parts of your brain light up. If you've got a good relationship, some of that activity is probably in your anterior cingulate cortex—it makes you feel warm and fuzzy. If there's been trauma in your life related to your mom—if she was abusive or passed away in a way that devastated you—then your amygdala probably become actives when you think of her. Your experiences have conditioned your brain to respond to different ideas and memories with appropriate feelings. If you wanted, you could recondition your brain to respond differently—but it would take lots time and work.

Here's why: Brains are energy hogs, and it would require enormous energy to ramp up your prefrontal cortex and your amygdala at the same time. This is one reason we think of rational people as cold, and why people often seem unreasonable when they're angry or sad. It takes too much energy to be analytical and angry at the same time. Your brain doesn't have the resources to pull it off.

There's a saying in neuroscience: "Neurons that fire together, wire together." The reverse is also true: The connections among neurons weaken when the activity among them slows down.

I didn't know this at the time, but all those hours I spent

researching and analyzing God were beginning to rewire the neurological network that made up my faith. If we think back to our brains as a large business or political organization, all my reading and studying was caused layoffs in the offices and departments of those whose job it was to make me feel the presence of God.

Every day, a few more employees were let go, resulting in more empty offices with the lights turned off. Before the president knew it, there was no one left in the entire organization helping me know God.

I spent so much time analyzing God that I didn't have time to experience Him. Over time, this caused the feelings I had about God to fade. God was becoming nothing more than an abstract idea, and one that could easily be torn away by the thinkers' books I was consuming. The process of losing God took months of reconditioning, but when the loss finally hit me, it felt like a heart attack: sudden and violent.

About 18 months after that fateful family meeting with my parents, Jenny and the girls were out of town.

I'm a family man, which means that I take no joy in an empty home. The first night of peace and quiet is usually nice, but before long, I want to hear the kids running and playing again; I want to feel the gentle rhythm of Jenny's breath as she sleeps. So when my family is away, I distract myself by catching up on things I can't do when my family is home. I watch artsy, esoteric films or horror movies. I read for multi-hour stretches, locked in the kind of fugue state that all bookworms crave.

On this particular night, I settled in with Carl Sagan's *Pale*

Blue Dot: A Vision of the Human Future in Space. The vast scope and scientific mysteries of cosmology were the frontier of my search for a better understanding of God. Cosmologists, astrophysicists, and astronomers generally aren't a religious bunch; they wrestle with the big questions of life in a tangible way.

By then, my faith was running on fumes. My prayers had degenerated into desperate pleas for some kind of sign that God was real, and I was afraid I'd never find my way back to a working understanding of Him. Scientific certainty was my security blanket now—a warm, albeit thin, blanket protecting me from a cold world.

Pale Blue Dot is a lovely book full of scientific insights and musings, and Sagan communicates it all with prose that borders on poetry. The book takes its name from a famous picture of Earth that was taken 3.7 billion miles away by the *Voyager 1* spacecraft. When *Voyager* was past Saturn—hurtling toward interstellar space at nearly 40,000 miles per hour—Sagan convinced the rest of his team at NASA to capture an image of Earth from that distant vantage point.

The photo transmitted by *Voyager* home to Earth is one of the most humbling images in human history. Our world appears as a tiny blue speck, incredibly small and insignificant in the vastness of black space. Of the image, Sagan said:

> Look again at that dot. That's here. That's home. That's us. On it everyone you love, everyone you know, everyone you ever heard of, every human being who ever was, lived out their lives. The aggregate of our joy and suffering, thousands of confident religions, ideologies, and economic doctrines, every hunter and forager, every hero

> and coward, every creator and destroyer of civilization, every king and peasant, every young couple in love, every mother and father, hopeful child, inventor and explorer, every teacher of morals, every corrupt politician, every "superstar," every "supreme leader," every saint and sinner in the history of our species lived there—on a mote of dust suspended in a sunbeam.

Sagan's words wrecked me. Nothing ever had shifted my perception of reality so violently. The universe is unspeakably vast and our planet an immeasurably small point in our solar system. Our solar system itself is just a speck of the Milky Way galaxy, which is just one member of a Local Group of galaxies.

And we think we're so important?

"For God so loved the world," seems absurd from the *Voyager*'s vantage point. Earth is a waterlogged pebble, one planet among countless others. What possible significance does the salvation of humankind hold?

I have those quotes from *Pale Blue Dot* in my journal. My notes further read, "This broke me, maybe forever." (By now, you know I can be melodramatic.) But I kept reading.

In the next chapter, Sagan's widow, Ann Druyan, an esteemed writer on cosmology herself, proposes an experiment:

> Look back again at the pale blue dot of Earth. Take a good long look at it. Stare at the dot for any length of time, and then try to convince yourself that God created the whole universe for one of the 10 million or so species of life that inhabit that speck of dust. Now take it a step further: Imagine that everything was made just for a single shade

of that species, or a gender, or an ethnic or religious subdivision. If this doesn't strike you as unlikely, pick another dot. Imagine it to be inhabited by a different form of intelligent life. They, too, cherish the notion of a God who has created everything for their benefit. How seriously do you take their claim?

After reading Druyan's words, I had to close the book. God was no longer merely distant. He seemed implausible. A myth concocted by frightened apes in a dangerous world. The most grandiose images in Scripture seemed laughably small when viewed from the perspective of cosmology.

Our God was a God of firmament and waters, of flickers of light in the night sky that seem somewhat insignificant when we stand on Earth and look up. But that view is backward. Amid the celestial giants of outer space, Earth is the one that's tiny and meaningless.

What miracle is parting a sea in the face of a supernova? What power lies in an empty tomb when you compare it to nebulae that are hundreds of light-years in diameter, giving birth to stars?

I went to bed anxious and fearful, unsure about everything I'd always "known." Sagan's words had found footholds in my mind where Dawkins's, Hitchens's, and Harris's had not. Sagan didn't bother with a direct assault on the absurdity of God. Rather, he shifted the frame and revealed the notion of a God who cares for our Earth as a silly notion in a cosmos whose scale dwarfs our imagination.

The morning after I read *Pale Blue Dot,* I spent most of my prayer time asking for insight—even pleading for a sign that God was real and could hear my prayers.

I pray every morning.

Prayer has been a part of my morning routine for as far back as I can remember.

Often I pray aloud. Somehow it makes God seem closer.

So that morning, I said these words: "God, I don't know why I'm praying. You aren't even real."

Two things happened immediately. First, the feeling I associated with the presence of God left me, like morning mist burned away by the heat of the sun. Second, I felt as if a trapdoor opened beneath me, and I fell through it.

A series of dark insights entered my mind with terrifying speed. I realized that all the people I had loved and who had gone before me were gone. Forever. I'd never see them again, because the only thing waiting on the other side of death was infinite blackness and the annihilation of self.

There was no heaven. There was no hell. Beyond this life, there was . . . nothing.

Without God, life had no objective purpose. All those tough days I had pushed through, believing that I served a higher purpose—that purpose was nothing but a comforting self-delusion. My life was meaningless. So were the lives of my children and of every person who had ever lived or ever would. It didn't matter what kind of husband or father I was, because all my hard work would be erased when the sun exploded in four or five billion years.

I felt a profound grief, an inky-black darkness, as I realized there was neither mission nor redemption for humanity. The universe was indifferent to us. We were all just an accident of the self-organizing principles of physics—mere quirks of gravity, electromagnetism, and chemistry. This was it. This was the end of my search.

"God, I don't know why I'm praying. You aren't even real."

In the time it took to say those 11 words, I'd become an existential nihilist.

And my parents got divorced anyway.

CHAPTER 4

Secret Agent Man

Tears came first.

Many Christians believe that people become atheists because they're angry at God. My experience didn't involve anger at all. I felt grief—a gnawing sense of loss. I felt empty and alone. God hadn't been some controlling sky king to me. God had been the embodiment of love, who cared for humanity with unknowable depth—and, by extension, cared for me, as well.

God had been my hope. God was the magnet that moved my moral compass. He was my Redeemer and my friend. I probably talked to God more than I talked to anyone else. After all, God was always there.

When I learned God wasn't real, it was like hearing that a dear friend had died unexpectedly. I smashed into this insight like a skydiver whose parachute refused to open.

What would happen on Sunday? I had a Sunday-school class to teach. I couldn't just show up and say, "Hey, guys, I don't think God is real anymore." That would certainly make for a lively discussion, but what would the consequences be? I couldn't stomach sending anyone down the road I was on. It was too painful.

In the eyes of my church, I realized, I was now a lost soul, an unbeliever. Once I told the congregation that I didn't believe, they would want to know why. They'd want to witness to me, lead me back to Christ. But what could they say that I didn't already know? I'd been studying the Bible and apologetics with the fervor of a man trapped in a well. What would happen when I had a retort for everything they said? They'd either shun me or start to walk the road of doubt themselves.

What about Jenny? Jenny's faith wasn't like mine—her convictions were deeply emotional, free of the logical scaffold that had failed my faith. What if she lost her faith because of me?

And what about our girls? Madison was starting to ask questions about God—the kind of questions Evangelical parents look forward to, because they mean a "profession of faith" is coming soon. Macey was too young to ask those kinds of questions, but she was starting to get into nightly prayers and talking to God. If I told my girls I didn't believe in God, their faith might never recover.

It wasn't just Jenny and the girls I worried about, either. It was also my mom. My dad, too. It was most of my extended family, including a beloved uncle who's a Baptist preacher. An overwhelming majority of my friends and family were Christians, specifically conservative Evangelicals. What would they think when I told them I was no longer one of them?

I realized I was in an incredibly vulnerable position. To profess atheism would be social suicide. My community couldn't tolerate it, and I could lose everything that mattered to me. Not because my friends were bad people—they weren't—but because this was the inevitable outcome of the Evangelical belief system. The flock must be protected from wolves.

So I did the only thing I could do to stay in the flock. I put on sheep's clothing.

I may have not been a Christian anymore, but I knew how to *act* like one. I'd have to lie, but that was a small price if it meant protecting my community and getting to hold on to the life I loved.

I loved my church. I loved getting up on Sunday mornings and spending a couple of hours with warm, friendly people. These were the same people I shared life with throughout the week. They were the friends I met for lunch, the people I saw movies with. Their children were my children's playmates.

But our church wasn't just a social club. The Southern Baptist Convention is one of the largest disaster-relief agencies in the world. Baptists take part of their collective offerings and set them aside for disaster relief. When the worst happens, Baptists are often on the scene with food, water, shovels, and trucks right alongside the Red Cross.

Our church was active on a local level, too. Several times a year, we collected food and filled up trucks to deliver it to low-income families. For all the talk about churches being insular organizations, I took great pride in how much sacrificial work my Baptist church did. What's a little lying to stay a part of all that?

It was not as if I'd forgotten what the Bible taught or what Baptist theology claimed. I knew more about those things than ever before. All those hours spent digging into the Bible to find its faults could now be repurposed, I hoped, to help people find happiness. The Bible was full of passages describing the importance of caring for the poor, giving generously, forgiving others, and being people of peace. I'd just skip over the genocide, proselytizing, and all that stuff about a fire that always burns and a worm that never dies.

My Sunday-school lessons became more powerful than ever. I spent hours researching the language and historical context for every week's Scripture passage and finding ways to connect those ancient words to actions we could take today to make the world a better place. I said prayers with people in times of distress. I talked to students about their doubts and helped them find confidence in their faith. I said a blessing at meals and led my kids in prayers when I tucked them in at night.

Holding the facade during worship services was harder. I love to play the bass, but the lyrics to praise songs now seemed meaningless—much like singing "Here Comes Santa Claus" all year long. Sermons were a bit easier to take, because our pastor was more likely to teach about being a good husband than he was to expound on the mystery of the Trinity. But I still had to fight the urge to cough, "bullshit," when his messages drifted toward the supernatural.

I was the world's least interesting secret agent—an atheist under deep cover in the Baptist church.

My duplicity was so deep that, one night, as an atheist, I actually led Madison to Christ. Honestly, that memory is too powerful, too bittersweet to offer up on these pages. But I will tell you that I thought it was better for her to believe something that was untrue yet comforting than to follow in my footsteps.

I wept when she was baptized. I was so proud and yet so ashamed.

I was able to hold on to my community, able to pretend to be a Christian with aplomb, but the psychological costs were high. My ruse was isolating and lonely, and I lived in constant fear of discovery. I couldn't allow anyone to know who I really was.

My closest "friends" became the authors of the books on science and philosophy I was reading—but those relationships were one-sided. Christopher Hitchens was unlikely to return my phone calls. And although I had developed a fascination with Carl Sagan in particular, I felt a perpetual melancholy that I'd only started to listen to him after his death.

I needed real friends who would understand me, so I started lurking around on Internet message boards where atheists congregated to talk about the challenges of being an unbeliever in the deeply religious United States. These boards were full of people who'd been abandoned by their families, whose marriages had frayed and failed, and whose employment options were limited because they'd gone public with their unbelief.

I didn't post anything—I was too scared someone would figure out who I was and expose me. But I felt a stronger sense of connection reading these conversations than I did while simply reading books.

More Americans than ever don't identify with any particular religion, but there still exists a tremendous stigma to being an atheist. Many Americans tend to view atheists as immoral, dishonest, and even dangerous. This couldn't be further from the truth.

Studies have shown that atheists are generally more moral than believers. Some studies have shown that the happiest, most successful children grow up in atheist households.

I grew up believing that atheists were angry at God. But listening to actual atheists taught me that most atheists simply want to be free to believe as they wish, without persecution from the faithful. Despite the media narrative around the New Atheists, most atheists don't even care if others believe or not. They just want to be left alone.

I am an extrovert of the highest order. I scarcely have an inner monologue at all—my brain is oriented toward putting thoughts out into the world, and I can only listen to a conversation for so long before I have to speak. So, more than six months after becoming an observer on atheist message boards, I decided to start posting as an active participant—but not without first thinking about how to cover my tracks.

The Internet is not as anonymous as it seems—it only takes a bit of technically savvy to puzzle out someone's identity. So before I posted anything, I created a set of aliases across multiple websites. I registered using multiple email addresses with various providers and never reused them. I had more online handles than a deck has cards, and I kept a spreadsheet of what account should post where and in what context. I didn't talk about where I was from or how old I was. I didn't want to leave any bread crumbs for the curious. If all that sounds crazy to you, you've never kept a secret that could cost you the things and people you care about the most.

Once I had my aliases in hand, I started to post. Questions popped out of my accounts like bubbles from a bottle of Coke that's been shaken. It wasn't that I needed more evidence against God. My faith was done. The problem was that I didn't know what to do now.

Ever since I'd confessed that God wasn't real, I had struggled to find a reason to get out of bed every morning. Humans are storytelling animals. Human consciousness constantly constructs a narrative wherein that particular human is the protagonist. In our brains, we're the good guys in a struggle against conflict and discomfort.

It's not enough for our physical needs to be met or for us to find material comfort. We need to feel that our lives carry some greater significance, that we make some contribution to the human drama. Even though humans are born with an intense survival drive, research has shown that we will readily sacrifice our lives if we believe our death has meaning. We'll override one of our most powerful instincts if it means there will be a powerful ending to our story before the credits roll.

This is the mechanism that drives soldiers to commit acts of heroism, firemen to rush into burning buildings, and terrorists to pilot airplanes into skyscrapers. We are desperate for meaning. And I'd lost mine completely.

My first set of posts were questions about how to find meaning without God.

"I don't believe in God anymore," I wrote. "In a lot of ways that's freeing. I understand that I can be moral without God, and that treating others the way I wish to be treated creates a society that is better for everyone. But why does it matter? No matter what kind of husband and father I am, no matter what I do, one day our Sun will swell into a red giant and destroy Earth.

"Let's say humanity gets its act together, and our civilization spreads to other planets and then other solar systems. There's a dozen ideas in physics about the eventual death of the universe.

We'll fall into entropic heat death, or protons will decay, or a new Big Bang will consume our observable universe as another is born.

"Any of these things will erase not only my contributions to good, but everyone's. There is no eternal significance to anything we do. This insight makes it hard for me to get out of bed in the morning or to enjoy life. My beautiful daughters are just like me—wood not yet turned into ash."

I cast a half dozen variations of that dilemma across the Internet. People took stabs at answering my questions, but nothing gave me hope that I could find a satisfying life without God.

About a week later, I got a reply of a different sort. This one was hopeful and helpful. It's been lost—the Internet's memory is imperfect. But here's my best recollection of what it said:

"I'm sorry you're going through this. A lot of us have been there, too. Some people cast off God as easily as a pair of jeans that don't fit anymore, but for others it's more like a death in the family. I'm sorry for your loss.

"You're finding that atheism isn't a belief system. It's just a lack of belief in God. You can't build a life on that. It's like trying to build a league of people who don't play golf. Total nonsense.

"But let's look at things with new eyes. Was the sunrise any less beautiful today just because it won't be around forever? Is the time with your family worthless because you'll die one day? Is life any less of a gift just because it's a result of physics?

"So, God gave you meaning. Do you still care about the needy? Do you still want to be a good father? Then do those things, make them your life's purpose. You don't need some God to tell you to be good—you can be good on your own. And isn't that more meaningful? To love and to make the world a better place because you choose to?

"You don't need God to make a purpose for your life. You can make a purpose for your own life—any purpose you choose. There's no angry man in the sky to smite you for making the wrong choices, and no savior to bail you out if you screw it up.

"You get one life, one shot to find every beautiful sight, to help others, and to enjoy the odd series of events that allow a bag of organic molecules to know they exist. Don't waste it."

I can't overstate how freeing it was to read this. It may seem simple, but it was transformational to learn that I could make my own meaning in life. This philosophy is called Secular Humanism, and when I was a kid in Youth Group, it was one of the bogeymen used to scare us into reading our Bibles. But in the search for meaning that followed my loss of faith, I discovered humanism to be a beautiful movement. Humanists see value in humanity; they see our species as full of potential and goodness. They seek solutions to life's problems of suffering and need, concerned only with improving the conditions of human life.

Humanism helped me salvage bits and pieces of my faith, such as those parts of the Bible that emphasized taking care of the weak and needy. Humanism made me feel more comfortable pretending to be a Christian on Sundays. Although humanists seek to solve problems through rational thought instead of religion, it seemed rational to me that religion was a particularly powerful way to motivate religious people to do good in the world.

I did have trouble breaking my daily prayer habit—even if my "prayers" were nothing more than vocalizations of my incredulity about the idea of God and my laments that there was no deity at all. On dark days, I doubted my doubt. I wondered if there was something I'd missed. I told God I'd believe again if I got a sign, even though I felt foolish and small as the words left my mouth.

Most of all, I wondered if I really needed to stay camouflaged forever. Could there be a way to admit what I believed while holding on to my family and friends?

I didn't know, so I asked the Internet. Using a website called reddit, I wrote a post that included more personal details than anything I'd posted in the past. I asked what losing my faith would mean to me, my family, and my community. I was terrified that I'd get found out by doing this, so I softened my language. Instead of telling the truth—that I was an atheist pretending to be a Christian—I said I was a Christian on the verge of atheism. That way, I would have an out in the event that someone read the thread and figured out it was by me.

I didn't have to wait long for answers. I had targeted a forum where both Christians and atheists hung out (although atheists have an overwhelming numerical advantage on reddit), and the post made it to reddit's "front page" in a few minutes. Posts that make it to the front page get a lot of views, so I went from tens to hundreds to over a thousand replies quickly.

As I read the replies, I noticed a few trends. Atheists tended to be empathetic. Many told me that if faith kept me happy, I should just keep on believing. I had expected to get a bunch of arguments against God, but what I got was shared humanity.

The Christians were not as gracious, as a group. Some were very kind and helpful, but most were vicious. They told me I would burn in hell, that I was going to destroy my family. They accused me of posting for fame and attention. Reading those posts four years later, I'm still surprised by the intensity of their anger.

This was all the data I needed to decide that I had to keep hiding. I could be myself online, anonymously. But in person, I had to maintain the facade, or everyone would turn on me.

And I did for almost two years.

CHAPTER 5

Love Seat Confessional

I've got a great marriage. I'm sure people get tired of hearing about it, but it's true. I married my best friend, but I don't call her that anymore. *Best friend* is too weak a label for what we have. *Wife* is better but still incomplete.

Jenny is strong where I am weak, and the opposite is often true. We make a great team, and there's no one whose company I enjoy more. I do get annoyed when she stops me on the way out the door to ask if I have my wallet and cell phone, but that annoyance evaporates when I realize my wallet is still on the nightstand.

Despite her keen sensitivity, I was able to avoid raising any suspicion about my loss of faith. I led the family in prayer, read my Bible, and even led my oldest daughter to Christ. I was diligent about concealing my unbelief—any sign of strain could break my cover. I'd fooled her so well, for so long, that it became second nature. I'd stopped worrying about Jenny catching on.

But that's when she did. To this day, I don't know what clued her in, but one night, she hit me with a subtle interrogation after the kids went to bed. We were sitting on our love seat, watching something on TV, when she turned to me and asked, "Are you OK?"

Jenny doesn't often ask me if anything is wrong. Somewhere in my mind, a warning switch was flipped, moving me from DEFCON 4 to DEFCON 3. How could she know? *Did* she know? If I didn't want her to find out, I knew I had to look casual.

"Oh, yeah, I'm good," I said.

"Are you sure?" she said. "It seems like something is wrong."

"What? No. I'm fine, really. I'm just tired."

"Is it me?"

Jenny said this with vulnerability, ready to own that some unintentional behavior of hers was bringing me down. This triggered my instinct to protect her.

"No! God, no, it's not you," I said.

"But it's something!" she said. She had me. I had let my guard down and admitted that something was wrong. It's one of the oldest tricks in the book, and I fall for it every time.

"Look, love, it's no big deal. It's nothing."

"Please talk to me about it. You seem distant."

DEFCON 2

A bead of sweat rolled down my back. Could she tell how anxious I was? I knew I was in a precarious spot—once Jenny catches wind of something, she follows the trail with the determination of a bloodhound. I tried to deflect, defuse, and dismantle the discussion, but she kept pulling the thread, kept digging deeper.

I had to give her something. What if I just told her the truth but did it casually? I relaxed my shoulders and tried to project an air of calm assurance before coming clean. "I'm just not sure that God is real anymore," I said.

Bad idea. Her eyes went wide, and her jaw dropped in a near comical look of surprise. She would have been less shocked if I'd slapped her.

DEFCON 1

Her cheeks flushed red, and her eyes were bright as she replied as only a southern woman can, "Well, you'd just better get right again." There was a budding anger in her voice, sure, but something more powerful as well. Fear.

Instinctively, she started to evangelize to me, relying on a lifetime of Evangelical training. She took me through relevant verses in the Bible that spoke of man's needs of salvation, the reality of hell, and the goodness of God's love. I was proud of her but also dismayed. I knew those verses better than she did. I had a stack of retorts for everything she said, but I would sooner kick a puppy than push against her faith. So I sat and listened, nodded, and smiled. I used my body language to communicate that she was doing a great job.

After a while, she was done. I hadn't said anything, so she said, "So?"

"'So,' what?"

"Are we good?" Jenny asked.

"Oh, yeah, we're good."

"You believe in God again?"

I had to be honest. She wouldn't fall for a lie. "No, not at all," I said.

And this kicked off a strange debate. She made a case for God, and I sat silent. She would then tell me my silence wasn't allowed. She wanted to know what I thought. I would offer the absolute minimum to answer truthfully and move the conversation forward. I didn't want to wreck her faith, but I could see how frustrated she was.

Even my halfhearted defense was enough to wear her down. I watched as the realization slowly dawned on her: This wasn't going to be solved that night. Her shoulders slumped, and her voice grew softer. Jenny always says I'm her rock, the stable ground she finds footing on, but I could see that she didn't trust that foundation anymore. She felt scared and alone. It was heartbreaking.

Atheism is making great strides in the West, and one if its most effective tools is rational debate. Prominent scientists make easy work of theologians, and atheists are well versed in taking apart religious rhetoric. Skeptics often know the theology and claims of Scripture better than rank-and-file believers do.

On the other hand, most believers aren't well versed in the claims of science, atheism, or skepticism. Plainly stated, atheists tend to win arguments, and humans enjoy winning. I knew plenty of atheists online who took great joy in breaking down religious people, dismantling their God into little LEGO bricks—a great monument to children's toys.

I did not have that zeal. It broke my heart to see good people lose God.

Jenny agreed that we should keep this a secret. She, too, could imagine the chaos that would come if people at church learned about my doubt. We were together on that.

We went to bed, and for the first time in our marriage, an iron curtain ran down the center of the mattress. We were in bed together, but we weren't *together.* A rift had opened up between us.

I usually fall asleep in less than a minute, but I didn't that night. I was afraid of what might happen to my marriage. I was also afraid that I had started Jenny down the road to atheism.

The next day, we went through the motions of feeding the kids and getting them to school. I went to work, I came home, and we put the kids to bed. Then we sat back down on our love seat. She didn't wait long before she said it.

"I'm not sure we can be married anymore."

When I tell this story in person, some people gasp. This is especially true for people who've never been members of a theologically conservative Christian church. But Jenny's not some monster, and she wasn't trying to manipulate me. Let me explain.

Jenny and I grew up believing that hell was a real place and that anyone who didn't have faith in Jesus Christ went there for eternity. While I didn't believe that anymore—my belief in heaven and hell fell away with my belief in God—Jenny still held that conviction.

In this worldview, if I prevented the girls from accepting the gift of salvation, they would spend eternity in hell. From that vantage point, her leaving me would be less an act of cruelty and more like a painful sacrifice to protect our children from the worst fate imaginable: eternal damnation.

Plus, we'd always been taught there was no love without God and that people who weren't Christians weren't capable of real love. When I said, "I don't believe in God anymore," Jenny heard a subtext: "I don't love you anymore."

But I was ready. I'd rehearsed my response for months.

"I've been an atheist for two years," I said. "Have I treated you differently?"

She said I had not.

"I love you. I love you more than I ever have. And you are no longer second place to an imaginary God—I can truly say that I love you more than anything. I love our family. I was an atheist when I lead Madison to Christ, and I'll do the same for Macey. I can maintain this act forever if I need to. No one will ever know I'm an atheist. I'll teach Sunday school. I'll be a deacon. Nothing has to change. We never have to talk about this again. I will do anything to make you happy, because I love you more than anything."

The imaginary-God part was overkill, but it worked. Her posture softened. She told me she didn't want me to have to live a lie, but she believed me when I said I really loved her.

We ended the conversation and went to bed—an uneasy truce for the sake of marital stability.

Jenny kept her word by keeping my secret. For one week.

I'd been alone with my private atheism for a long time, and you'd think this shared secret would've created some kind of solidarity between Jenny and me. It didn't. We were alone together. She couldn't talk to me about her grief and loss, because I was the source of it all, but she couldn't talk to her friends, either. It was too much to bear.

When Jenny broke her silence, she called in the big guns. She talked to my mom.

Jenny's tactics were brilliant. The only person I know who

prays more than my mom is my grandmother—her mother. Somehow, my mom reads her Bible 12 hours a day and prays 14 hours a day. My mom has the profound, unshakeable assurance of someone who actually believes God is sovereign and cares about the minutiae of our lives.

The news of the day doesn't shake her, but her faith doesn't make her indifferent to the suffering of others. Mom reads the Prophets in the Old Testament, and she cares for the poor, the orphaned, and the widowed. Mom is the reason why, even though I let go of God, I could never accept that religion was bad for humanity.

My mom is both religious and a profoundly good human. She was the perfect person to call in for reinforcements. And she blindsided me when I least expected it.

Letting go of God had an unexpected benefit: I got better at science. It's much easier to learn about quantum physics, infinite cosmic inflation, or the boundary of a black hole when you're not trying to shoehorn it all through a God-shaped lens.

I honestly felt as if I'd taken a drug that boosted my IQ. Concepts I'd struggled with for years "clicked" with sudden clarity. But this delightful cognitive boost was offset by an emotional numbness. I'm not just talking about loneliness or depression here. I'm talking about something unique to people who've lost their faith: transcendence withdrawal.

Don't try to google that. I just made it up. Now, let me explain it.

Many people who believe in God have profound experiences when they pray or worship. They feel a sense that God is near them and cares deeply for them, and this sense can sweep the believer into an emotional epiphany that has a euphoric quality—a high so intense, it could come from a drug. You may have seen this on the faces of Christians who raise their hands when they worship.

If you've never had one of these experiences, you have no idea how powerful they are. Unlike the high achieved from an illicit substance, soaring religious transcendence carries no hangover, no rough crash from a high place. The only lingering effect is seeing the world as more loving, recognizing the love God has for every person.

But if transcendence had been my drug, astronomy was my methadone. When church lost its meaning, my cathedral became the night sky, my chosen worship instrument a telescope. It wasn't as emotionally powerful as worshiping God, but my spine tingled more than once when I looked through the lens and considered that the photons striking my retina had left some star thousands or millions of years ago.

If you've never studied astronomy, you may not realize that the sky is a time machine. Light moves really fast, but things in space are incredibly far away, and 186,000 miles per second doesn't cover much ground on a cosmic scale. When you look at the moon, you see the moon as it was 1.3 seconds ago; when you look at the Sun, you are looking eight minutes into the past. Saturn, as we see it, appears as it was about 80 minutes into the past.

The scale gets even bigger when we leave our solar system. When I look at Proxima Centauri, one of the stars nearest our own, I roll the calendar back 4.2 years. If I turn my gaze at Orion's belt, I see its stars as they were 736, 1,340, and 915 years ago, long before America was a nation. Andromeda is the closest major galaxy to our Milky Way, but looking at it takes you back 2.5 million years—long before humans appeared on Earth.

The farthest we've looked back is 13.8 billion years, to a murky haze of microwave radiation that gives us a "baby picture" of the universe when it was only 380,000 years old.

If that doesn't amaze you, check your pulse: You may be dead.

My telescope was more than just a scientific instrument. It was my means for communing with the universe, for relating to objects that were trillions and trillions of miles away. Some of these objects were already gone, ghosts in the sky whose final brilliance hadn't reached our planet yet. But here I was, witnessing their beauty, all while standing on Earth.

I didn't know it at the time, but there's a good reason that astronomy and vast, outdoor spaces yield an experience similar to intense religious contemplation. Brain scientists have found that the farther you shift your focus away from yourself and toward an expansive view, the more likely you are to feel this kind of awe.

The believer focuses on God, and his mind's eye is drawn beyond himself, to his family, his friends, his church, his community, his nation, his world, and to God—a vantage point more expansive than any other. The astronomer stands in a vast, open space and casts his gaze into the cosmos, across distances that make the might of Mount Everest seem absurdly small.

Both the Christian and the astronomer feel a drive to pass along that feeling. For churches, this means proselytizing; for astronomers, it means throwing star parties.

Star parties are every bit as nerdy as they sound. A bunch of people, mostly older men, gather in a field near dusk and set up telescopes. For some reason, they do so in a rough oval, with their cars parked as close as possible to minimize the distance over which they have to lug their gear. There are tables for star charts and snacks, and strange cultural norms prevail. (For example, any light source must be red, because white light kills your night vision. You can always tell a first-timer because he'll inevitably turn on a flashlight—to the audible dismay of the group.)

It's a lot of work, and the best sights often don't show up until the wee hours of the morning. But people do it because it's great fun to share this view of the night sky with others. Most people have never looked through a telescope, and some of the objects out there are stunning. When viewed with the naked eye, Saturn is a yellow point of light in the sky. But seen through a telescope, Saturn will take your breath away. The rings are textured, and the orb itself looks almost as if it's made of soft fabric. People seeing it for the first time often think they've been tricked; they walk around to the front of the telescope to check for a sticker. After all, how could that little yellowish point turn into something so grand?

Two weeks after telling Jenny I was an atheist, I hosted a star party for my church friends in a big field down the street from my house. I sent out a Facebook invite, telling folks to come out and bring their kids. I must have done a good job with the invitation, because a lot of people came.

The star party made me realize how close I was to blowing my cover. Everything was going fine early in the night, when we were looking at the planets and nearby stars. I was explaining to everyone how the sky was a time machine, and when I found a star 2,000 light-years away, someone commented that the light had left its star around the time when Jesus walked the earth.

Oohs and *ahhs* all around.

But the next stop on our interstellar journey was Andromeda—the major galaxy closest to the Milky Way. One of the kids asked, "How long ago are we looking now?" I answered, "Two and a half million years." Before I could invite everyone to imagine what was happening on Earth that long ago, a staff member from church spoke up. "That's not possible. The universe hasn't been around for more than 10,000 years."

Somewhere, a needle skipped across a record. You could have cut the awkward silence with a knife.

I didn't have the heart to start an Old Earth versus Young Earth debate, so I just said that some apologists believe God created the universe with the appearance of age—that part of "Let there be light" may have included some light waves already en route.

During this exchange, I noticed that my mom was watching me closely. Mom can be a night owl, so I wasn't surprised she had hung around for as long as she did. But I started to wonder if she was there to check out more than the stars.

I continued pointing out different highlights in the night sky as long as people wanted, but many started drifting off toward home, and by two in the morning, only Mom and I remained. I was trying to home in on Neptune or Uranus when Mom worked up the nerve to say what was on her mind.

"So, Michael, I talked to Jenny . . ."

Before she finished the sentence, I knew exactly where this was going. Suddenly, Mom's newfound persistence as an amateur astronomer made sense. Like a surgeon who has spotted a tumor, Mom set to work diagnosing my doubt, looking for what needed to get cut out for the patient to live.

I was reticent. Mom's faith was pretty unshakable, but I had seen the man behind the curtain. I knew the true folly of faith. I knew that any reasonable person would be swayed if she knew what I knew.

I told Mom I didn't want to hurt her faith, but that's sort of like telling a mother bear you don't want to hurt her cubs. "Don't worry about that," she said. "You aren't going to hurt my faith. God and I are tight." So I did with Mom what I couldn't bring myself to do with Jenny. I spoke openly and honestly about my doubts. I told her why I thought no reasonable person could believe in God if he considered the evidence.

It was a sparring match not unlike the one between Agent Smith and Neo in *The Matrix*. "Theism versus atheism—only one can survive." Only I didn't feel like a liberator. I felt like a jerk. With utter dispassion, I made a counterpoint for every one of her points, then a counter for every counter.

Hours into our talk, Mom lost her fire. With a defeated tone, she said, "Well, I guess you have an answer for everything." I told her that I missed God and that I wanted to believe. I told her about how I used to cry out to God and ask for a sign. Then I told her how all I got back was a deafening silence, a void that made me understand that if there was a God, that God was so distant as to be meaningless to us.

Mom said the devil had a hold on me. I told Mom the devil didn't exist. Mom quoted C. S. Lewis.

Finally, the sky started to take on a purple hue—the sun had started turning night into day. I told Mom that I loved her, but I was tired. Mom said she loved me and then left me with one final parting blow:

"Michael, I am going to pray that God will move so powerfully in your life that you can't deny it's Him. Hold on."

Some skeptics are offended when people offer to pray for them. I never was. Sure, sometimes "I'll pray for you" is a passive-aggressive quip. But more often, when someone says she'll pray for you, she's truly saying, "I care for you deeply, and I think about you a lot. I'm going to ask the most powerful force in the universe to help you."

Even if there's no God at all, if a believer prays for you, it means she cares. So I thanked Mom for caring, even though I felt bad for her. Miracles did not really occur. I knew her prayer would not be answered.

What can I say? I'm a sucker for misplaced confidence.

CHAPTER 6

NASA and Bacon Numbers

In the world of nerdery, there are elites, just as in any other culture. Elite nerds are well versed in computers, science, or graphic novels, and they work at places such as Apple or Google or even Marvel Comics.

But if nerdom has an ultimate pinnacle, it would have to be NASA. Apple and Google have globe-spanning computer systems, and Marvel creates unforgettable imaginary worlds. But NASA put people on the moon and landed an SUV-size, nuclear-powered robot on Mars. NASA's nerds are the "elitest" of the elite. They demonstrate the power of mathematics, science, and engineering in a way few other endeavors can.

You can imagine how excited I was, then, when I found an email from NASA in my inbox.

As I'd shifted more and more energy toward learning about astronomy and space exploration, my social-media profile started to reflect that passion. This shift is what set me up to get an email from NASA. NASA is a big government agency with more than

18,000 civil employees and over 40,000 contractors, but its most famous projects loom so large in our culture that the vast majority of its other work gets buried under photos of *Curiosity* or the International Space Station. In hopes of getting the word out about research being done at their Dryden Flight Research Center, NASA was writing to tell me it was opening the center's doors to the public for the first time and inviting a handful of bloggers to document the festivities.

They told me my mission was simple: Take pictures, ask questions, and post my thoughts online. (As if it's possible to take a VIP tour of NASA and not want to tell other people about it.)

I booked a flight and hotel stay and added a couple of extra days to tour nearby Los Angeles. Though I'd been there on business, I'd never had time to see anything there other than conference rooms. Los Angeles is one of America's most famous cities, and I looked forward to finding out what all the fuss was about.

A week later, my old ad-agency compatriot Stratton Glaze emailed me. By this point, Stratton had moved to Los Angeles for seminary, and he had also gotten a job working for Rob Bell, the author and pastor who had prompted my move into Progressive Christianity. Stratton told me Rob was planning a small, 50-person conference on creativity. Stratton was working with the team that managed Rob's website, and they needed some help testing the store. If I helped them, I'd have a guaranteed spot at the conference—my ticket would be locked down before the public got access.

I was torn. Not about helping Stratton, but about going to the conference. I'd reached a point where I was comfortable and confident in my unbelief. I didn't pray anymore, and I could imagine a day when I didn't go to church anymore. I wanted to be a humanist without the baggage of pretending to be a Christian. This seemed like a step backward.

But there was good reason for me to go. My career had changed. I'd started out in IT, but a shift toward digital in the ad business was pulling nerds out of the server room and into the boardroom. I was in charge of the new-media teams at our company, which meant I now played an active role in the creative process of coming up with ideas and campaigns for clients.

It was fun work. Where else could I get paid to put the Aflac Duck on Facebook or make a website that ages people's faces to show the effects of smoking? But it was also terrifying. I never knew where my ideas came from, and sometimes they wouldn't come until the last minute. I lived with a constant fear that my most recent idea was also my last.

I talked to Jenny about it. I told her about my worries about creativity and how this conference was supposed to address them. Rob Bell was a pastor, sure, but he was also the author of multiple best-selling books, and he was known for giving talks that left an impression on people. I'd heard that Bell was a master at building storytelling suspense, even as he broke down complex, ancient ideas about God. I didn't care about explaining God to people, but I needed to figure out how to explain insurance in a 30-second video on Facebook, so I wanted those secrets.

Jenny asked when the conference was, and I told her it was right after my NASA trip. I would already be in Los Angeles, so I didn't even need to book an extra flight. I saw this as a pleasant coincidence, but Jenny couldn't contain her enthusiasm. She leapt at the idea of me sitting in a room full of pastors and religious folks for two days. My mom was even more excited. She kept telling me to "just hold on, because God is gonna move."

I tried to be gracious, but it annoyed me how much significance my wife and mother placed on this timing. What they saw as the hand of God was nothing more than pattern-matching systems in human brains.

But they were right. As it turned out, that trip would leave one hell of a pattern on me.

NASA was everything I'd hoped for and more. The Dryden—now Armstrong—Flight Center is part of Edward's Air Force Base. It requires a long drive through the desert to get to the gate, followed by a long drive even farther into the desert to reach the actual base. Older experimental planes are on display all over the place, starting before you even get to the main gate and then dispersed every few miles on your way in. I actually shouted when I pulled into the Dryden parking lot and saw an honest-to-God SR-71 Blackbird on display in the sunlight. To me, standing next to that dark machine was like meeting a movie star.

Inside the center, we saw amazing things: Global Hawk drones that spent their time studying climate and weather, software that prevented airplanes from crashing due to pilot error, planes that were designed to fly quietly at supersonic speeds. We saw research aircraft, and we were buzzed by a fighter jet right before another passed overhead to give us the experience of hearing a sonic boom.

At one point in the tour, our guides stopped us and told us that we couldn't turn around. Some secret prototype aircraft was rolling out of a hanger for an engine test. We spent a few minutes standing there, listening to something roar.

I wanted to turn around and look, but I remembered the waiver I'd signed that said deadly force was authorized on the base to protect classified information.

The amazing technology wasn't the best part. Neither was sitting in a fighter jet or seeing gear used to test the Apollo mission. What I enjoyed most was the community of nerds. At Dryden, I wasn't a deacon. I wasn't a Christian at all. No one here cared that I thought the universe was billions of years old. No one asked about my thoughts on God. We were just a bunch of science nerds hanging out on the frontier where great science is done.

For the first time, I felt that I was able to be myself—my true self. I took off my Christian mask and felt the sunlight on my face.

There's a study that says 42 percent of Americans will undergo a faith transition at some point in their lives. They'll leave the tradition they are a part of and move on. This statistic must mean that a lot of people in churches are wearing masks. They feel alone, but, really, they aren't. There are others wearing masks, too. Behind those disguises, they just can't find one another.

Our churches will never be healthy as long as those experiencing doubt feel they have to hide. In too many churches, the response to doubt and tough questions is shaming, passive rejection, or probes about a possible "sin problem" in the questioner. All this reaction does is push the doubting away from their faith. It has to stop.

Or that 42 percent will never come back.

At the end of a long day in the desert spent looking at the pinnacle of humanity's technological prowess, I sat in a tiny hotel room in the middle of nowhere and reflected on my time at NASA. I felt a warm afterglow from finally being who I was. It's not that I'd walked in and proclaimed I was an atheist—I hadn't. NASA was just free of Christian language and culture.

We'd had a dinner the night before the tour of Dryden, and someone did ask me what my faith was. I told her, "I don't have one." She said that she didn't, either, and the conversation moved on without a hitch. I thought about how nice it would be to experience that kind of freedom in my life back home.

The next day I drove into Los Angeles, where my friend and coworker Colleen was celebrating her 25th birthday at a bar downtown. By some odd coincidence, many of our close friends were all in LA at the same time, so it was a birthday party/reunion of sorts.

I'm a big guy: 6'1" and over 230 pounds. Despite that frame, I'm a lightweight when it comes to alcohol. I didn't even drink until after age 30, so I get a solid buzz on one beer. I get drunk on two, and I have trouble walking after the third. I don't drink often, so when I do, my friends are always waiting for "three-beer Mike."

Because we were all celebrating, I had my three beers—and a couple of other drinks, as well. I was in that state where you move like a balloon in a breeze: floating, yes, but also bumping into things.

Our crowd decided to find a place that was less congested. Once there, my friends wisely switched me to liquids that would hydrate but not inebriate. It got late, and the party finally began to wind down. But then the birthday girl had a brilliant idea: karaoke. Everyone except me groaned. Our gang had more than its fair share of introverts, but I'm always down for karaoke. My boldness is matched only by the tone deafness of my singing voice.

We staggered out onto the street in search of a musical thrill, and the sounds of drunken, communal singing nearby spilled onto the sidewalk. Transfixed, Colleen and I followed the sound to its source, with our friends following grudgingly.

We found a small bar packed with people. There was a microphone, yes, but it was late enough that the songs being chosen were old, familiar standards, and everyone was singing along. We joined the chorus seconds after walking through the door.

A couple of songs in, I looked around the room. Everyone was smiling and singing. They were strangers an hour ago, but now they had their arms around one another's shoulders. Another song came on, far out of my range, but I sang it anyway.

So did everyone else. The person with the mic was tone-deaf, but it didn't matter. Her voice was just one sound in a cacophonic symphony. People held hands. They sang without self-consciousness or shame. They weren't performing, either. In all my years of singing in pews, I'd never heard anything so beautiful.

Somehow, in this small bar in Los Angeles, people had left their egos behind and were simply celebrating life together. There was no shame or pride. There was no posturing. There was just the music and the joy of being alive. As the song built to its larger-than-life chorus, I started to question the way I looked at the world. The source of the easygoing, upbeat dispositions in that bar was easy to identify: the inebriation associated with our metabolic response to alcohol. But understanding that chemical origin didn't explain everything I sensed in that room. There was some quality to that moment that my evidence-only approach to the world couldn't articulate.

As the song hit its highest point, and the bass rattled my bones, I thought of my daughters. When they were small, they'd reach up to take my hand anytime we crossed a street. Their tiny fingers would wrap around mine.

You can describe this via materialism. An unfathomable volume of atoms move toward one another. When they get close enough, electron boundaries repel one another. None of the atoms "touch," but an exchange of virtual particles creates a field that can't be crossed without some crushing force such as the gravity in the center of a star. These atoms are all rattling around because of thermal energy, and they radiate infrared photons as they do.

At a higher level, these atoms form molecules and then cells and then tissues. Nerve cells detect the tactile pressure against skin cells and the heat radiated by the flesh of another body's hand. This signal is converted into electrical impulses that travel up the arm and into the spinal column.

Once there, these signals are sent to multiple parts of the brain and mix with other signals, too: The nose sends up samples of molecules that the brain calls *shampoo.* The visual cortex turns a torrent of photons into the shape of a little girl's head and her tiny, fine strands of hair. And so more ancient parts of the brain light up, both electrically and with the release of chemicals: neurotransmitters and hormones.

A father feels love for his daughter and wants to protect her. You can explain it all with science—and with stunning accuracy.

But that explanation can't tell you what's it's like to be an actual father who loves his daughter, who would do anything to protect her and keep her safe. Explaining the physics, the chemistry, the biology, and the neuroscience of this moment is like projecting sheet music onto a movie screen instead of listening to the symphony.

To explain the love I have for my daughter, I can't ask the scientist for help. I need the poet or the painter. I need a song or a sonnet. Beauty can only be described with beauty—there's no substitute.

In that bar with my friends and the singing crowd, I felt the presence of something greater than even the grandeur of the cosmos, something more mysterious the teeming subatomic particles that compose our reality. I searched my mind for a word that would describe this perfect moment. It was more than beautiful, more than sacred.

Holy.

It was holy.

I'll never be able to listen to "Don't Stop Believin'" again without shedding a tear.

Have you ever played "six degrees of Kevin Bacon?" It's a simple game. You pick an actress or actor and see how quickly you can connect them to Kevin Bacon via on-screen appearances. Matt Damon is two degrees from Kevin Bacon: he and Casey Affleck appeared in *Good Will Hunting;* Casey Affleck and Kevin Bacon appeared in *Lemon Sky.*

Therefore, Matt Damon's Bacon Number is 2. So is mine.

Not that I'm much of an actor. I've only been in one movie: *Blue Like Jazz,* a film adaptation of the book with the same name. Marshall Allman stars in the film, playing a young Don Miller, and I appear on-screen with him as an extra. He appeared alongside Kevin Bacon in *Jayne Mansfield's Car.*

How I got into the movie is a long story, but I'll give you the gist. *Blue Like Jazz* was a really helpful book for me at a difficult time in my faith, so when I found out they wanted to make a film of it via Kickstarter, I pulled out my credit card and pledged what I could afford—even though my faith wasn't in a great place. Later, the film's creators offered donors the chance to appear as extras in a couple of scenes.

If you don't believe me, just go watch the trailer for *Blue Like Jazz*. There's a scene where a cross piñata is lowered in a church, and I'm sitting in the front row.

Don't judge me. I weighed almost 300 pounds in that film. I was busy coping with the fact that I didn't believe in God but was appearing in a Christian film.

I woke up early the day after our karaoke party. One of the benefits of being a big guy who's a lightweight is that I never drink enough to have hangovers. But the rest of our crew wasn't so fortunate, so I found a nice breakfast spot and after my meal went for a long walk.

Our larger party was down to a smaller set: Stratton, Jonathan, Colleen, and me. That put me in a very paternal mind-set. I'd mentored all of them at some point in their careers. Stratton and Jonathan had both moved on to new gigs, but Colleen was still working with me.

They are three of my favorite humans. I don't think of them like my kids, but I do think of them as people I'd like my kids to turn out like.

By some quirk, *Blue Like Jazz* was playing nearby. I'd never seen it and wanted to, so I asked if anyone wanted to go with me.

Stratton had a class at seminary, but Jonathan and Colleen were game. It was just the three of us in the theater. The lights went down, the film started, and in no time at all, I was looking at myself on a movie screen.

I couldn't stop laughing. It's so surreal to see yourself projected onto a wall, like a mountain troll in a blazer. Jonathan and Colleen laughed, too, though I think my facial expressions on-screen had a lot to do with that.

But my mood shifted when I saw a change they'd made to the movie plot versus the one I knew from the book. In my scene, young Don gets angry and storms out of the church. He's angry because is his mom is sleeping with his youth pastor, a man who is also his mentor. Don is so hurt by this betrayal that he goes from being a good Baptist to an angry rebel, pulling his faith apart piece by piece.

Somewhere in my brain, a pipe became over-pressurized and burst. I don't know why, but my thoughts turned to the God I no longer believed in, and anger rushed through me. I realized that God had failed me just as He'd failed Don Miller. This God whom I'd loved and worshiped, whom I'd trusted, who was supposed to be all-powerful, had sat by and done jackshit while my parents' marriage fell apart.

Where was God when Dad started to fall in love with another woman? Was he distracted? Where was "the One who can part the sea" when the two of them held hands for the first time? Where was the Almighty when my Mom prayed for things to get better? "Great is Thy faithfulness"? Hardly.

Great is thy neglect, great is thy cosmic indifference. Great is thy absence.

The anger didn't last long. It soon turned into sorrow, an aching sense of loss over what was and what could have been. I thought about going fishing with my dad, of saying prayers with my mom, and laughing around the dinner table. I thought about the wholeness of a family that was centered around God.

A God who never was. No wonder we fell apart.

I cried during the whole damn movie. I cried as Don walked further and further away from faith, as he pushed away those who had been closest to him. I cried as Don owned up to his role in the mess and the way he'd hurt others.

For the first time, I realized that my deconstruction of faith hadn't been the rational and clinical pursuit I believed it to be. When I opened Genesis, I wasn't just looking for answers: I had a bone to pick. I'd wanted answers for Dad, sure, but I'd also expected answers for myself. I'd expected God to justify Himself to me, but God had failed to do that.

Instead, He bowed His head and died.

I couldn't explain to Jonathan and Colleen why I was crying so much. When I tried, I only sobbed louder. They're good friends, so they just drove me around and chatted with each other while I tried to put Humpty Dumpty back together.

I had to get to Laguna Beach for the conference, so Jonathan and Colleen decided to drive in and stay overnight, too, which would give us a day to knock around town. There are worse places to spend a free day. As we drove from Pasadena to Laguna Beach, I gathered myself, slowly drifting from inconsolable back to my normal disposition.

We got in late, after dark. Jonathan and I decided to run

out into the ocean and were surprised that the undertow was so strong. We were exhausted when we got back out of the water, laughing the laughs of men who've escaped from unexpected danger.

The next day was wonderful. We had a relaxed day in a beautiful beach town, eating good food and drinking great wine. Late in the day, Colleen told me she'd accepted a job with Google, which made me all the more afraid for this day to end. It was the NASA effect: I was enjoying being myself, without my Baptist mask.

All good things, however, must come to an end. Flights must be caught. Jobs must be attended to. Jonathan and Colleen drove away, and I took time to reflect on how I was changing. I'd grieved a lot in the past few days, but it was so liberating to stop pretending that I believed in God for the first time in my life.

I could feel a page turning, a page that marked the end of that chapter in my life. I wondered what waited for me in the next one.

CHAPTER 7

The Horse Leaves the Barn

The next morning, I came into the conference like a starship prepared for battle: shields up, on red alert. My experience at NASA had opened my eyes completely to something important: I was tired of hiding. I was tired of pretending to be a part of the grand fantasy called Christianity. How could I be otherwise?

Spend enough time hiding who you are, and it'll warp you. My life had become a body-snatcher film. On the outside, I looked like any of the other people at church. I sang the same songs, I prayed the same prayers. I wore the same clothes, and I bowed my head when everyone else did.

But on the inside, I was different. I was pretending. Where they had a God-shaped hole to fill with worship, I had the conviction that God was a comforting fairy tale told by apes smart enough to know they'll die one day. My carefully composed Christian facade was peeling away. That weekend, I had admitted my doubts to friends, both religious and not. I just couldn't keep up the act anymore. I'd started to fantasize that Jenny would leave faith behind, too, and we could slip away from church, free to enjoy our limited time as conscious entities alive and aware together.

Free of delusions. Truly free.

As I approached the little bungalow that hosted the conference, Rob Bell was standing outside to greet people. He didn't look anything like the guy on the book jackets. He was taller than I expected and a lot more tan. He looked as California as they come. But seeing him made me realize that I really didn't want to meet Rob Bell or anyone else. All I wanted to do was mine some secrets about creativity and then get the hell out.

So I slipped past the small receiving line and evaded the people already inside. One perk of my early arrival was that I found a spot on a sofa in a room full of folding chairs. If I had to be a pod person, at least I could be one with a comfortable backside.

People continued to filter in and find seats, and before long the room resembled a can of hip, sharply dressed sardines. It was a lovely space, even with the elbow-to-elbow occupancy. Natural light streamed in from French doors that faced the Pacific, and the "stage" consisted of a stool and a large poster pad with markers.

Rob bounded in and got the show under way. He opened by describing the emotional fatigue people often feel when they have to do creative work as part of their occupation. He talked about creativity and the rituals and life patterns that made it easier to come up with ideas and story lines without falling into mental exhaustion.

Then he started to talk about human consciousness and how

it develops—genuinely fascinating stuff that wasn't religious in nature. I didn't feel like a pod person anymore. This group seemed far less prone to magical thinking than did my church back home.

Around mid-morning, someone asked a question that evoked René Descartes's famous maxim, "I think, therefore I am." From there the group talked about rationalism and the Enlightenment.

Even more surprising to me than the high-minded turn in conversation was how accurate the discussion about science and philosophy was. I'd never heard Christians speak with such knowledge about secular ideas. Rob and these other Christians had a genuine appreciation for the good things that science—and a respect for empirical evidence—bring to society.

But then someone mentioned miracles, and my "Kumbaya" moment unraveled.

Rob said, "Science is based on repeatability, but the problem with that is that, by definition, miracles aren't repeatable. Science has nothing to test, so science can't handle miracles."

I thought: "No. Science is based on recorded observation. We can't repeat the Big Bang; we merely observe evidence that indicates it happened. Likewise, any 'miracle' would leave a mark on the physical world that we could test and observe."

Rob talked about how the repeated success of science in explaining things drove the notion that *every*thing can be explained. This notion, he said, drives New Atheism, "which sounds a lot like faith to me."

My hair stood on end.

I thought, "I am an atheist, and I've never heard anyone say that."

I could feel the red warmth rising in my cheeks. I hate angry outbursts, so I did what I usually do when I feel anger coming on I channeled it into sadness and helplessness, just as Rob was bringing up the subject of holism.

Rob described holism—the theory that says the parts of a whole are interconnected to such an extent that they cannot exist independently of the whole—using human consciousness as an example.

"Where does it come from? Atoms make cells, cells make tissues, tissues make organs, but where are *you* in that?" Rob said. "Where is the person who likes Mexican food and hates mayonnaise? We can't find you in your elbow. There's a 'you' that can't be located in your elbow."

I felt a weight on my shoulders and a tightness in my belly. I realized I had come here for more than creative insight. Deep down, I was hoping to find some new angle that could make God real to me. In this moment, I realized how foolish that hope had been.

God wasn't just dead. God had never existed in the first place.

We can't find "you" in your elbow for a reason. Your body is the support system for your brain, and your brain is where "you" are. You can lose your arms and still be you. Your eyes can fail. You can get a new heart. But your brain is the part of you that is uniquely you. The pattern of neurons, synapses, and dendrites in your skull is demonstrably where your thoughts, feelings, and actions originate. Muck around with gray matter, and you can change a person.

If some external, immaterial spirit is the real us, why is it that our brains drive everything we do? We may not understand consciousness precisely, but observation makes it clear where it—and we—comes from: the brain.

Rob and the group moved on to the subject of evolution. I

think most people in the room accepted evolution as fact. At one point, Rob said, "Evolution does a lovely job explaining why you don't have a tail, but it cannot tell you why you find that fact interesting."

I hung my head. Literally.

Actually, evolution does quite well at explaining why humans find things interesting.

Natural selection gifted us with a neocortex that works with our hippocampus to create curiosity. Novel stimuli attract us, and that creates learning. Our ability to learn from diverse experiences is one of the things that led us to global dominance as a species.

The group continued to talk about how science tells us how and faith tells us why.

I'd heard that phrase more times that I could count. It even has a name: non-overlapping magisteria, or NOMA for short.

My eyes actually started to well up with tears. Every time Rob would raise some new point that challenged the ideas promoted by skeptics in the New Atheism movement, the energy in the room soared. The Internet is full of "true stories" in which some Christian high school or college student humiliates an overconfident science teacher or philosophy professor. Even in this roomful of basically science-literate believers, I saw the same need to create an enemy who could be defeated. But instead of dealing with the real claims and implications of empiricism, they dealt in truisms and sound bites.

I needed to speak up, but I didn't want to. We southerners hate to make a fuss, and I don't like to challenge people. But these Christians actually spent time with people who didn't attend church. If they were to have any effect on the world, they needed to understand the shortcomings of their approach toward unbelief.

Rob continued his train of thought about holism and how it changed our understanding of reality. He said it was obvious that the universe is growing in complexity, and . . .

I couldn't stay silent anymore.

Like a nervous second-grader, I raised my hand.

Rob looked at me and said, "Yes?"

Now, I've been speaking to crowds my entire life. Churches, trade shows, boardrooms, whatever; I don't get nervous. But this was different. In the moments before I spoke, my heart galloped around my rib cage, and the room seemed to shrink.

My voice shook as I began to speak.

"Look, I really don't want to say anything, but I will never have an opportunity to be in a room with Rob Bell and a bunch of cutting-edge Christian leaders again," I said. (Always start with a compliment.) "I've spent the last two years going to a Southern Baptist church as a closeted atheist, and I had a moment when I doubted my doubt after I saw *Blue Like Jazz* yesterday—a movie that I am in, if this story could be any more complicated."

People erupted into friendly laughter.

"Wait, you're in the movie?" Rob said with genuine bemusement.

"I'm in the movie, yeah," I replied.

Rob didn't miss a beat. "So you're a Southern Baptist atheist?"

"I'm a Southern Baptist atheist who's been an atheist for two years but hasn't told anyone," I said. "How do you like that story?"

I offered my best self-effacing smile. I broadcasted my uncertainty as much as I could with my body language. I used every trick I have in my arsenal to deal with authority figures when I feel threatened.

Rob responded, "Well, that horse is out of the barn."

Everyone started laughing again. I get less nervous when people laugh. I start to feel more comfortable. So I continued:

"Here's the point. I'm struggling, because I've always been a science/technology/math/engineering person—which is a lot of fun as a Baptist. But I discovered that my faith had to be destroyed so that I can live."

Rob's gaze was like a laser, and he nodded. "Yes?"

I kept going: "I was so with you on everything we discussed earlier. Everything about human development and consciousness. The idea of just loving people where they are is really affirming to me. I go to a church of loving, sweet Fundamentalists."

Laughter echoed across the room again.

"No, really, they are. They're like, 'We love gay people. We're going to change them, but we love them,'" I deadpanned.

I knew I was on a roll, because people were laughing more than ever, so I went for the throat: "My problem is, there are all these Christian people that I love and respect. But even in this room, where comfortability with science seems much higher than I'm accustomed to, I don't think you get it. Science can explain beauty. In fact, I was talking to a friend who's an astrophysicist, and he thinks theologians fear the next stage of human consciousness, which is unbelief. He thinks that even loving, Progressive Christians are fearful of what a post-God world looks like. They can't picture it.

"But a post-God world doesn't mean that love isn't real," I continued. "Scientists can certainly describe and understand love and beauty and why humans are so compelled by them. There are things that may be unknowable in science, such as what exists beyond the temporal edge of our universe, or even what's beyond the edge of the visible universe. That edge is moving away from us at the speed of light, and it's redshifting as it does (a redshift is the way the wavelength of light coming from objects moving away from an observer seem to grow longer). We have no way of making observations beyond the temporal edge of the universe, and we may not ever have one. But just because we don't know what's beyond that edge doesn't mean that God is real."

Now the room was very quiet.

No one was moving in his or her seat. It was so silent that people might have been holding their breath. Only the waves of the Pacific continued to audibly inhale and exhale, indifferent to our conversation.

"These are the things I'm stuck with," I said. "I'm a person who wants to believe on some level. It's the reason I cry out to God, 'Are you still there? Jesus, are you His son?' I want to believe that need is something intrinsic to me, but actually I believe that this need for God is just a lot of social conditioning."

Now it was so quiet, I was afraid I'd upset and made people angry. So I did what works for me in such situations: I dug in deep and spoke the most complimentary truth I could find.

"I apologize for prevailing on everyone, but I realized if I let this moment pass, it would never come again. I'm in a roomful of people who know so much more about this stuff than I do."

A few let go of nervous laughs, so I closed with: "How can anyone who understands how the universe works believe in God?"

That horse really was out of the barn.

I expected to be run out of the conference on a rail. People can be vicious when you threaten their deeply held beliefs, and I had just told a roomful of pastors and religious folk that no reasonable person could believe in God.

But Rob leaned toward me in his chair and said something unexpected. "Thank you. On behalf of everyone here, thank you. I think we all needed to hear that."

A few people started to clap, and when I looked around the room, I didn't see angry faces. As far as I could tell, folks around me were concerned for me and my struggle. No one looked ready to correct or scold me.

Rob took a moment to collect his thoughts before speaking to me again directly.

"Somewhere within you is a longing, and if I were you, I'd try not to overanalyze it," he began. "I might not go much more beyond the fact you have this longing, and so far you can't seem to find a scientific or other category to meet that longing. It's as if, in your life you have ideas, and 'this fits in this bucket, and this other thing fits in this other bucket,' but you have this other thing that is real but doesn't fit in any of your buckets.

"I would call that the God bucket."

Rob looked at the ceiling for a second, his right leg bouncing up and down—either nervous or excited—as he thought about what he wanted to say.

"I probably wouldn't define it much more beyond that," he said. "Your mind, which is obviously seriously dialed in, is probably used to mastering things. But it sounds like, in your experience, there is this thing you cannot master. It's happening *to* you. You aren't standing over it in a white lab coat, holding a clipboard. This thing somehow does something to you that's different from anything else. It's almost as if there is inductive reasoning, deductive reasoning, and *ab*duct-ive reasoning—which is the stuff that kidnaps you. You know what I mean?"

Easy laughter returned to the room.

Again, Rob sat back. "You're sort of saying, 'This thing abducted me, and I don't know why. It is somehow real, but I can't get myself to it through the standard ways that I know. But it is real,'" he said, his eyes beginning to shine. "I get sort of choked up. I think it's beautiful you're here. So, I assume coming here was some odd step of faith."

I told him that it was, and how I'd just been at NASA, looking at what mankind can accomplish through science.

Rob smiled and said, "You know, it's interesting. A lot of people here were at NASA on Friday."

This time I was the one laughing, so Rob continued. "You are here, and there is something in you that doesn't go away even when you become an atheist. I say, let's all celebrate that. There's no need to define it further—our words will just screw it up. I think that God, if there is a God, doesn't ask you for anything more than that. I really believe that God is that which we can't stop talking about, and that God is what happens when our words fail.

"Both of those things happen at the same time," he said. "You just told me that you don't tell anyone about your doubts, because you don't want to hurt their faith. That's sacred and beautiful. You're already living a Jesus life, so let's just celebrate that."

When Rob talked about the "God bucket," I felt an electrical current run through my body, as if someone had turned a dial in my skull's control room.

For a brief second, I felt the presence of God again. It was powerful and fresh, but it died out just as quickly as it had appeared.

Then this big bald guy with pierced ears spoke up (his name is Bradley, and he's since become one of my closest friends).

"Brené Brown says that the opposite of faith is not doubt. Faith and doubt need each other. The opposite of faith is certainty," Bradley said. "When I heard that, I realized, no wonder I was such a screwed-up Fundamentalist. But when I let the doubt just be there, my faith grew."

Murmurs of affirmation erupted around the room. A corporate executive named Sarah stood up and talked about meeting God on top of a mountain in Nepal, outside of any church context. Her story was riveting and dogma free. It was unlike any testimony I'd heard before.

She wasn't selling the idea of Jesus. She spoke of him like an old friend.

I realized that the people in that beach house that day accepted me exactly as I was. I didn't feel like an outsider. I didn't fear being found out.

I threw the fullness of my doubt about God at them, and they held it with grace. They didn't shout me down or take apart my arguments. They didn't try to win me over or rebuke me.

They just accepted me. And they even thanked me for caring.

When I look back at that moment in that room with fifty strangers, I imagine what would've happened if someone had rebuked me, told me that the devil was after me, or told me I was in rebellion against God. The people in that room simply accepted me, wholly and completely. But I realized they weren't the first who had done so.

Despite all her fears and anxiety about my doubts, Jenny had stuck with me. Instead of drawing away from me, she had leaned in closer.

My mother had taken the time to make a case for God, but she'd done so in the context of love. Our relationship as mother and son was never used as a weapon, and I was never threatened with maternal exile.

If you're a Christian who wonders what to do with someone who's in doubt, consider these words carefully: Love and grace speak loudly. The first and best response to someone whose faith is unraveling is a hug. Apologetics aren't helpful. Neither are Scripture references. The first thing a hurting person needs is to know they're not alone.

My path back to God was paved with grace by those who received my doubt in love.

Let's return to that moment when I felt the spirit of God again. I'd known that feeling my whole life, and I've never quite been able to put it into words. Somehow, it's a feeling of God's presence, a numinous sense of proximity, like a personal version of the "Spirit hovering over the waters" in Genesis. I hadn't felt that sensation since I told God that I didn't believe anymore, but here it was again.

I found that confusing and disorienting. It's not as if I believed again—I still had the same rational objections to the existence of a supernatural deity. It's just that when I pondered Rob Bell's challenge to put all my questions in a mental bucket called God, I felt God again.

After a couple of years of researching this phenomenon, I think I have an idea of what happened in that moment.

Anthropologist Tanya Luhrmann suggests that American Evangelicals train their brains' reality-sensing mechanism to project some of their own internal life onto an external source. In essence, the rituals and culture of many churches teach Christians to experience God personally by paying special attention to certain thoughts and feelings.

But how could I experience this so suddenly? It's not as if I went on an extended spiritual retreat or spent hours praying and trying to feel God again. My conversation with Rob Bell had lasted fewer than 10 minutes.

I think the answer lies in neurons.

People who believe and pray daily invest tremendous neurological real estate in maintaining a model for "God," and I had spent 30 years building and reinforcing this model. Researchers have found that people who build such elaborate neural networks around God use that network as part of the way they experience reality. These networks are sophisticated, and they connect critical parts of the brain related to attention, memory formation and recollection, language, and emotion.

Think back to my office-building analogy from earlier. By this time, my God network was a building full of empty offices and employees who had been laid off. When Rob Bell challenged me to contemplate the unknown, using the word *God,* it was as if some custodian flipped a circuit breaker down in the basement, and suddenly the lights came on in thousands of empty offices all at once.

My neurological CEO saw the sudden flood of light and was astounded. Why were the God offices full again? Who had authorized that activity? But before the CEO could pick up the phone, the unseen custodian flipped the breaker off, and the lights faded again.

For the rest of the conference, I was filled with hope. I wish I could say that I believed in God again, but that's not true. It's more that I felt there was some way to make peace with religion and to better understand what works about it.

I was reminded that humans have been surrounded by mystery and uncertainty for as long as we've been able to look around us and ask "why."

During the rest of the conference, Rob taught us more about creativity and how to craft stories and sermons. At the end of our second day together, we all had dinner together at the restaurant next door. We talked and laughed and drank a little. I felt so free. I didn't have to filter everything I said through a lens of biblical innerancy, Christ's atonement, or a God whose nature and motives I pretended to understand.

After dinner, we gathered one last time to end the conference with a ritual.

As I walked downstairs, I saw a small table in the center of the room, set with bread and wine. I knew immediately what was about to happen.

The Eucharist. Communion. The Lord's Supper.

It goes by many names but is practiced, re-enacted, and celebrated across Christian traditions.

Still, I thought it was a corny way to end what had been an amazing experience. It seemed an end more appropriate for a junior-high Bible camp than an intelligent, nuanced time of growth and innovation.

But I took my seat and crossed my arms as Rob began to talk about Jesus.

He said that the first time people did this, before they shared the bread and the wine, Jesus went around the room and washed the feet of his followers. And then he talked about quantum physics.

I unfolded my arms.

Rob said the bread and wine in front of us was ordinary matter. It was made of protons, neutrons, and electrons, and there was nothing special about it—it was simple matter.

But when we celebrate the Eucharist, we take ordinary matter, and we bless it. We make it holy and sacred, and we set it aside for a special purpose.

Rob paused to look around the room and then told us that we were all just ordinary matter. Like the bread and wine, we were made of protons, neutrons, and electrons. The Eucharist reminds us that we, too, can be set aside, be made holy and sacred.

We are set aside for a special purpose.

Like Jesus, we are to be broken and poured out for the healing of others.

That sounded beautiful to me. It sounded like humanism.

Rob said that if we didn't know what to pray, we should just ask God how we could be broken and poured out for others.

This next part sounds crazy—so crazy that for a long time, I omitted it from the story. I can't explain it, and I won't try. All I can tell you is, this is what I experienced. Make of that what you will.

Before coming to Laguna Beach for Rob Bell's event, I hadn't been praying much anymore. The only times I did was when I prayed with others to maintain my cover or felt a strong bout of nostalgia.

So I felt a little strange as I bowed my head and said the words Rob suggested.

I asked God how I could be broken and poured out for others.

Nothing happened. I didn't feel anything, other than a little foolish.

Who was I praying to?

God?

Who is God?

Is God even a who?

Am I praying to myself?

It was all very confusing.

People started to get up, one by one, and walk toward the front of the room where Rob, the bread, and the wine waited. He offered each person a piece of bread, and each would take it and dip it into the wine.

Some people exchanged parting words with Rob. A joke, a handshake, a hug. And then they would leave.

It seemed easy enough. I wasn't sure when to go up, so I waited for a break in the procession.

Rob's eyes got teary when he saw me walking up from the back of the room. That seemed silly to me. I didn't feel anything at all. In fact, I was completely numb, as if I were walking on a bunch of pillows. It was very strange.

When I finally got up there (it seemed to take a long time), Rob broke off a little piece of bread and held it out to me.

He said, "This is the body of Christ, broken for you."

Those words threw me into an existential vortex.

What? The body of Christ?

Jesus wasn't a real person. How could his body be broken?

Even if Jesus was a real person and really was crucified, how could he know who I was? We were separated by millennia, language, and progress unfathomable to a first-century rabbi who was convicted of treason. His body couldn't have been broken for me. That would be impossible.

And what would it help?

Are we saying God sent Himself in the form of His own Son and then died to protect me from Himself? The punishment He Himself was going to give me for not believing in Him?

That didn't make any sense.

The problem was that I couldn't take the bread without taking the metaphor. And that felt dishonest. I didn't believe that the body of Christ was broken for me, because I didn't believe that there was a body to break.

I decided to walk away. But just when I was about to turn, I heard a voice say, "I was here when you were eight, and I'm here now."

I froze, startled and amazed.

I thought about hiding from bullies and talking to Jesus.

I thought about growing taller and stronger and asking Jesus to forgive me for all the sins that I enjoyed so much.

I thought about my wife the first time I saw her in a wedding dress, and my oldest daughter as she went under the water at her baptism. I thought of the effortless charm my youngest daughter displays when she has the whole family laughing at the dinner table.

I thought about all the awful things I'd done even when I knew better, and yet how full of laughter and love my life was despite that.

I thought about my best friend, a Jewish rabbi whom I'd never met in person but talked to more than anyone else.

So I reached out took the bread from Rob's hand.

Then Rob said, "This is the blood of Christ, shed for you." I dipped the bread into the wine, and I ate it.

I took the bread and the metaphor, and I ran from the room, my face full of tears.

This is the part where I should explain the science of how a sane person can hear an audible voice in a room when no one has spoken. Believe me, I've spent a long time researching it, and I would love to explain it.

I can't.

The closest thing I can find in the sciences are hallucinations. Maybe that's what happened. Maybe I had so much longing and pent-up emotion that I fell into a semi-hypnotic state in a very suggestive environment. The bread, the wine, the prayer—there's a reason the Eucharist is a sacrament. This table has spoken to people confused about God for thousands of years before I picked up *The God Delusion.*

Even though it's my life's mission to help reconcile God and the sciences, that process breaks down at this point in my story. I can't explain what happened in that moment.

Which is unfortunate, because hearing Jesus speak to me was nothing compared to what happened next.

Late night bled into early morning, but I was too wound up to sleep. There was too much cognitive dissonance in my thoughts, too much unfinished business with God for me to call it a night.

So I walked down the steps at the back of the hotel to the beach. It was between 2:00 and 3:00 a.m., dark and still. I looked at the ocean, but it was so black, I couldn't tell where the water ended and the sky began—a powerful force that I could hear and feel but not see. As metaphors for God go, that was pretty good, so I faced the waves and started to pray.

I said something like this:

"God, I don't know who or what you are.

I don't know anything about you.

I don't know what your relationship to the Bible is.

I can't unlearn all the things that made me believe you aren't real.

They're still there, and they tell me you can't exist.

If you're real, if you have consciousness, will, and power, tell me:

How can I be here? Why did you bring me to California to learn about you from a famous preacher, while right now children are starving to death all over the world? Why answer my mom's prayer, when another mother's prayer that her child be spared from warlords goes unanswered?

God, that doesn't seem like love. That seems like evil. How can you just watch and do nothing? People are hurting and dying down here.

Look, I can't promise to accept the Bible now. I can't say I'll swear to keep its commandments. Some of them are archaic and brutal. Some of them are absurd anachronisms.

But I do like talking to you again. I feel you near me again, and I missed that, God. I don't want to be away from you anymore.

So let's make a deal. I will try to do the best I can to do good in this world. I will serve others, and I will work against suffering. But I have to keep asking these questions about your justice and mercy. And I can't forget about science.

Let's just keep talking about this, You and I. I don't ever want to be away from you again. I can't do that anymore.

All I know is, I met Jesus tonight."

When I said the word *Jesus,* the waves rushed toward me. I was standing high up on the beach, 25 feet or more above the waves, but the water still rushed up and over my feet—all the way up to my shins.

I thought about what Rob had said: that Christ's last act of service before his crucifixion was to wash the feet of his followers.

I said, "Is that you, God? Is this really happening?"

And the whole world fell away, like the veil lifted from the face of a bride on her wedding day.

Time stopped. The waves seemed to stand still, as if an unseen hand had pressed pause on the universe's remote.

Have you ever tried to look through a sheet? You know how if you stretch the sheet tightly, you get a hazy image of whatever's on the other side of it?

As I stood on the beach in the wee hours of the morning, everything in my surroundings took on that stretched, translucent quality. I could see what I can only call the glory of God on the other side.

I felt God with me, in me, and through me.

I felt connected to the Source of Life and the Source of All.

Through God, I felt connected to everyone else, all of humanity. And then, to all life on Earth.

All my doubts and questions were swept away in an ocean of light. I had no more pain or sorrow.

Suffering made sense as a part of a grand tapestry, one of the millions of colors in the palette of creation.

These words point toward what happened to me, but there are no words for my experience that night.

I left this world, and I went to another far more beautiful than this. And in that place I met my Maker face-to-face. Of course there was no face, but we are treading in places now where language becomes impotent.

I don't know how long it lasted, but it was by far the most powerful moment of my life. It was like my first kiss, my wedding day, and the birth of my children rolled into one moment and multiplied by a zillion.

Even as I type these words, I am moved to tears and a profound sense of awe, merely from the echo of that moment.

After it was over, I understood why someone would feel compelled to write about a bush that burned but was not consumed.

Or a blinding light on the road to Damascus.

Or an angel telling a 14-year-old virgin girl she was pregnant with the Son of God.

There weren't words to describe the things they, and I, experienced.

Scientists call this phenomenon a mystical experience.

Such experiences have been studied, documented, and analyzed. When researchers have asked people to lie in a PET scanner and describe such a moment, brain scans show that any attempt to put the experience into words will change the experience in your mind.

Mystical experiences transpire outside the realm of thought and language.

Mystics refuse to try to describe them. They simply sit with the experience and let it change them.

So that's what I did.

The next day, I had an appointment at our Los Angeles office, so I got into my rental car and started driving north. I talked with God the whole way there. It was like being on a road trip with a dear friend you hadn't seen in a long time.

We laughed, asked questions, and told stories.

It was as if God was physically *there* with me, in that rental car, just as present as He was in the Arc of the Covenant or the Holy of Holies in the Temple.

Yes, I am completely aware how crazy that sounds.

Yes, it's completely bonkers.

A lot of what God and I talked about that day was my confusion over God's existence. I had become an emotional and experiential Christian who also was an intellectual atheist.

Are You there, God? It's me, Michael.

There was a tension I couldn't resolve, no matter how much I tried. But recently I learned a little brain science that helped me make sense of it.

Your brain is divided into two hemispheres: left and right.

The two halves of your brain communicate via a thick channel of nerves called the corpus callosum. In men, the two halves of the brain don't talk to each other too often. They essentially exchange emails when absolutely necessary: "Hey, pick up the toothbrush; we have to brush our teeth."

This relative silence is one of the reasons men can focus so well. If my children are quiet in another room, they cease to exist in my mind.

Women, on the other hand, have much higher activity along their corpus callosum. The two halves of women's brains are in constant communication, and this allows my wife to be aware of everything, everywhere, all the time. Her consciousness creates and sustains a far more detailed map of our home than mine does.

Have you ever heard what happens when a microphone gets too close to a loudspeaker? There's an awful, high-pitched squeal called *feedback*. It's what happens when a sound system hears itself.

Feedback is self-amplifying and self-sustaining. In the 1960s, scientists learned that epileptic seizures are a form of feedback: feedback on the corpus callosum between the two hemispheres of the brain.

In the most severe cases, epileptic seizures can be life-threatening. Neurosurgeons once decided to try a radical procedure to treat these patients: They surgically severed the patient's corpus callosum. No one knew what to expect when the first patient awoke after surgery. Surprisingly, the patient seemed unaffected and recovered fully.

This was the case in multiple surgeries. Patients were cured of epilepsy but otherwise remained unaffected—or so doctors thought, until one of the patients punched his wife.

The guy didn't mean to hit her. He was trying to give his wife a hug, but when his right arm reached out for an embrace, his left arm threw a punch.

Another patient was picking out a dress for work, some sensible pattern, when her left arm grabbed a much louder print off the hanger and chose it over the dress in her right hand.

A third patient couldn't sleep. He was terrified that his now-unpredictable left hand would strangle him unless he stayed awake all night.

Imagine that.

This strange, unsettling phenomenon is called alien hand syndrome, or AHS, and it's unique to patients who've undergone this particular surgical procedure. Something happens when the two hemispheres of the brain can't communicate directly.

Scientists weren't sure what it was, but they had some ideas. The two halves of your brain are basically identical. If you really have to, you can survive with only half your brain. But there are some differences between the two halves.

For example, in most people, only the left half of your brain can initiate speech (the right half is mute). Not only are the hemispheres of the brain subtly different, they're also wired to the rest of the body backward. Your left hemisphere controls the right side of your body and vice-versa. So, when patients with AHS had a misbehaving left hand, this told scientists that something was going on in the right hemisphere of the brain.

The prefrontal cortex is the brain's decision maker. Scientists believe it's where all the competing loops and systems of the brain converge, creating the experience we call "consciousness."

Generally, the left prefrontal cortex is the dominant one, so, by some definitions, the "you" you think of as "you" is in your left prefrontal cortex.

Your consciousness may be associated with your left prefrontal cortex, but you also have a right prefrontal cortex. Almost every structure in your left brain is mirrored by one in the right. Scientists hypothesized that, perhaps without the corpus callosum, the left prefrontal cortex lost its influence over the right. Perhaps alien hand syndrome happened because the two hemispheres of the brain begin to "drift." Doctors even began to wonder if each of us carries around a fully-conscious right brain that serves as mute slave to the left.

So they devised an experiment that would allow them to communicate with each half of the brain in isolation. Both hemispheres of the brain can use both eyes, but they each only get half of your field of view. With a careful combination of special glasses, monitors, and positioning, the researchers could present questions in such a way that only half the patient's brain could see them.

There remained one small problem: The right brain can't speak. (Remember, there's no language center in the right temporal cortex.) Fortunately, scientists are clever. They decided to leave Scrabble tiles on the table, within easy reach of the left hand.

What follows can be described only as freaky.

One of the earliest experiments involved a young boy. Scientists flashed the word *girlfriend* where only his right brain could see it. His left hand spelled out the name of a girl, and the boy blushed with embarrassment. He knew the name was special, but he didn't know what question he'd been asked.

Physicist Dr. Michio Kaku interviewed neuroscientists about such experiments in his book *The Future of the Mind*. One subject in an experiment had been asked what he wanted to do after graduation. He said he wanted to be a draftsman, a very practical occupation. But when they asked his right brain, his left hand spelled out *automobile racer*.

Draftsman and automobile racer are wildly different occupations. One had come from the subject's rational, focused left brain, while the other came from his artistic and open right brain. The two brains had different agendas and goals for the future, but they were living in the same skull.

Another patient was asked what he believed about God. His left brain spoke and said he was an atheist, but his right brain said he was a believer.

An atheist and a believer co-existing in the same skull. Two halves of the same brain matter, flesh and blood.

So does half of a person's soul go to heaven and the other to hell?

Does Jesus live in only half of a person's heart?

Many of us can relate to this paradoxical experience.

Have you ever prayed fervently while simultaneously wondering if anyone was hearing that prayer?

Have you offered someone comfort in faith, while wondering if you believed anything you were saying?

For all its bizarreness, the phenomenon of split-brain patients gives me strange comfort. Suddenly, I don't feel so weird for identifying with both skeptical and spiritual people.

There is an atheist in my brain who remains wholly incredulous about the idea of a divine being who once dwelt among us in the form of a man.

There is a Christian in my brain who is indescribably and enduringly comforted by the idea and love of a supernatural Savior.

I've stopped trying to deny, starve, or otherwise do away with either of them.

I let my atheist question and examine. I let him check my motives and search for ideas that can be proven. Atheist Mike contemplates ethical issues from all angles, where right and wrong emerge not from ancient texts, but from the relation between our actions and the suffering or consent of others.

Christian Mike views the world through a lens of great compassion, seeing pieces of God in all His creations. My Christian side suffers with those in pain and finds reason for hope in everyone. Against all reason, Christian Mike believes it's never too late for redemption and that salvation is always at hand.

Christian Mike wants to drop his fishing net and follow Jesus. So I let him.

And Atheist Mike tags along for the ride.

PART II

GOD IN SCIENCE

CHAPTER 8

Happily Ever After?

I recently shared my story—the one you just read—onstage at a big church in Texas. This particular rendition included high-definition video, stage lighting with complex cues, a string ensemble, and vignettes interspersed with music and contemplative prayer exercises.

It's a dramatic and complex way to tell a story, and it always makes me nervous. During the presentation, I spend all my time offstage reading the cue sheet, because I'm terrified I'll walk on at the wrong time or say the wrong thing when I do. But on this night, I could hear people sniffling in the audience—a common reaction among people who've doubted or experienced a moment with God in some way.

Anytime I tell my story on a stage, I always hang around afterward to chat with audience members. Many come up to tell me how moving the story was to them or how it helped them make peace with a question or fear they've wrestled with. For people who've known God but struggle with some kind of doubt, my split-brain analogy is comforting. It helps them sit in the tension of their faith and their doubt.

Other people are less moved. They don't see how an ocean wave and a voice that no one other than I heard proves anything. They all ask me a variation of the same question: "How do you know all of that didn't just happen in your head?"

On this particular night, as audience members were waiting their turn to meet me, I noticed a man toward the end of the line, dressed in black, a shock of dirty-blond hair, his posture stiff, his fists clenched. I recognized him as a member of my tribe: the nerds. My suspicions of his nerdery were confirmed as he moved closer to me in the line. He was wearing a World of Warcraft T-shirt.

I wasn't sure if he was nervous or angry, but the closer he got, the more confident I became that the answer was both. His glower was hard to ignore, even as I listened to a woman tell me about how son had left the church after reading a book by Sam Harris. When our conversation was over, she hugged me and walked away, and the man in black stepped forward.

"Science Mike," he said, "you're either delusional or a fraud. I've listened to a lot of your interviews, read your 'doubt' series, and listened to your podcast, and I've never heard you say anything that makes me think believing in God is rational."

He took a deep breath before continuing. "You talk about brain scans, as if that proves anything. We can do a brain scan while someone thinks of a unicorn, but that doesn't make unicorns real. Or we could brain scan someone on LSD, but that doesn't make their hallucinations part of material reality. If you really heard a voice, don't you think a socially induced hallucination is the most likely source? And the thing with God in the waves—you'd had a few drinks, right? Couldn't it just be a moment of transcendence in your brain triggered by alcohol and a wide-open space? Isn't that much more likely than the Creator of the universe showing up for you while stubbornly obscuring any evidence that other people could actually use?

"Look," he said, "you have a right to believe whatever you want to believe, and you seem to be a kind and mostly reasonable person. My problem is, you're a smart guy, so people hear you believe in God and then decide to continue their own, completely unjustified pursuit of a faith that harbors pedophiles, oppresses gay and lesbian people, and demands that everyone else adhere to their ideas about morality. Christianity is a plague on humanity, and your story is an infection vector."

I smiled. "Thanks for having the courage to say all that," I said. And I meant it. "But if you follow my work, you know I'm not out to convince anyone of anything about God. My work is in response to suffering—there are people for whom the loss of God produces acute pain. Second, you're right about everything you've said. My experience doesn't prove anything to anyone—not even me."

My moment on that Laguna Beach shore isn't the end of the story. If my life was a work of fiction, that moment would have all the qualities of a story climax, but there was no "happily ever after," no credits that scrolled as I walked back to the hotel. Life isn't fiction, and that bright, unseen light was not the end of my story—it was the start of a new chapter in my life.

God returned to me in those waves, but I had no idea who or what God was. I didn't understand what the Bible had to do with God, or even if Jesus was a real person from history. I'd experienced something, sure. But the two halves of my brain didn't agree on what it was.

My inner Christian was affirmed and ready to proclaim this "Good News" with evangelistic fervor. But my inner skeptic was offended that "God" was even in the running as a possible explanation. If you found a $20 bill in your jacket pocket, you wouldn't assume Santa Claus put it there. How did I know my experience had anything to do with God?

In the next few chapters, I'm going to do what there's never time to do on a podcast or on a stage. I'm going to take you through what I discovered in my search to make sense of my experience on the beach and to relearn who God is.

We'll look for God in the earliest moments of the universe and in the laws of physics. Most religions say that God dwells within us, so we'll check the most likely place in our bodies for that to happen.

We'll look at the science of prayer and what studies tell us about prayer's effectiveness at changing our behavior and outlook.

We'll take a close look at the Bible and see what merit it may or may not hold in helping us learn about God.

Then we'll turn our attention to the Church—its historical and contemporary place in society—and the effects, both positive and negative, it has had on our culture. We'll look at the most puzzling question of all: Why Jesus? Why do I choose to follow a man who hasn't been seen in 2,000 years and who, some scholars argue, didn't even exist in the first place?

In these chapters, I want to show you how I've learned to let go of certainty. How I shift at will between different frames of viewing the world, depending on which is needed in each moment. This approach allows me to learn and follow science without having to surrender any line of evidence to suit my theology, but it also allows me to connect with God in worship and prayer—to know and contemplate that which seems impossible in the viewpoint of science.

Our Western culture wants a clear winner, and a lot of this has to do with our neurological craving for certainty. My prefrontal cortex wants me to believe that it's in control, that through it I'm making rational decisions based on an objective assessment of reality. This illusion is great for helping me sift though the overwhelming amount of information that is reality, but it's an oversimplified picture of what's really going on. Contrary to what we may feel, we humans weren't designed to find truth or objective reality. We were designed to find food and to mate, to avoid predation and to maintain social standing in our tribes.

This matters a lot if you're someone who's trying to reconcile your inner Christian with the part of you that feels skepticism and wants hard proof for the things you believe. Most people in this state struggle against it, their brain's need for certainty and social identity creating a contest to learn which one is "right." But studying neuroscience, physics, and cosmology has shown me that there are limitations on thinking that way, and those limitations are especially dramatic for anyone who wants to know God.

But I've learned that the need for certainty is an addiction we can kick—that it's possible to have faith, and even follow Christ, without needing to defend historical Christianity like a doctoral thesis. We can approach beliefs not as gems to be mined from the earth and protected with clenched fists, but as butterflies that land on an open hand—as gifts to enjoy but not possess.

CHAPTER 9

Einstein's God

By the time I met God on the beach, I had spent two years pretending to be a Christian for the sake of social stability. At church, I had become adept at hiding, in large part because the Internet gave me an outlet for expressing my true thoughts without fear of judgment or rejection.

But when I came back from California, my "secret agent" act stopped working. My Baptist friends understood my renewed need to know God, but they were really uncomfortable with my questions and the way I discussed my hang-ups with the Christian faith. The atheists I'd befriended online were all too happy to discuss the problems of Christianity, but they became uncomfortable to the point of hostility if I brought up my experience on the beach.

My friends online compared me to a mental patient or an alcoholic off the wagon. I was personally attacked not only on a forum thread but via direct messages. Voices that seemed reasonable and helpful when I was in doubt turned hostile and threatening the moment I expressed hope that a "God" of any description could be real.

Once again, it felt as if I no longer had a space where I could be myself. Both the Church and the Internet were uncomfortable with my search to know more about the God I'd met in Laguna Beach. So I mainly searched alone.

Around this time, my friend Cory Pitts called me—the student pastor who'd lent me Don Miller's book six years earlier. In the time since my parents' divorce, Cory had made his way to another church several states north. We didn't talk much anymore, so I was really happy to hear from him. He had a way of helping me make sense of life's murkier situations.

I told Cory about my time as an atheist and my return to a faith that felt much less defined. But I also told him about the nagging doubt that made me wonder if my experiences with God were made up. I told him how I prayed every day but wondered if I was just praying to my imagination; how singing worship songs or hymns could send me into a funk if the lyrics spoke too much of God's actions, of His superiority to other gods, or even of Jesus' sacrificial death. He mostly listened and asked a lot of questions, but I also remember him encouraging me to rest, because conflict like this can sap your energy more than you realize.

The next day, Cory sent me an email. He told me about a friend of his who'd been very sick. Her illness advanced to the point where she was hospitalized, wrecked not only by the original disease but also by infections that had opportunistically attacked her body. In time, she became so weak that her heart stopped and she had to be resuscitated.

Cory's friend was once vital and strong, but her illness had reduced her to a second infancy. In the hospital, others fed her and clothed her; she was completely dependent on them for survival. Over time, she had to relearn how to perform the most basic functions: walking, brushing her teeth, even feeding herself.

Cory said my faith was like that. My faith had died and been resuscitated. Like his friend after her heart was restarted, my faith was alive again but weak. If I wanted to remain a person of faith, I had to learn to do everything over again.

The first step on that long journey was to address the most basic question of all: How do I know God is real?

When someone says, "I believe in God," they're being vague. It's a necessity because when people discuss God, they're often working from a false assumption: that we all mean the same thing when we say *God.* It's a reasonable assumption. Most words that are used as commonly as *God* have meanings we all agree on. You don't see heated Facebook debates over the meaning of the word *chair* or what chairs mean to our lives today.

But God is a complicated idea, historically and philosophically, and there are an incredible number of definitions for who or what God is. Consider this:

There are atheists, who lack belief in any god or gods.

There are antitheists, who assert that belief in God is harmful.

There are agnostics, who say they don't know who or if God is.

There are pantheists, who say that the universe is God.

There are deists, who say God made the universe but that God does not intervene in the universe anymore.

There are nontheists, who find both atheism and theism too limiting. They believe God is real but beyond any human understanding or definition.

There are theists, who say that God is a being with specific will, agency, and a plan for humanity.

Even among theists, there are thousands of conflicting ideas about God. The world's three largest monotheistic religions all point toward the God of Abraham, but they disagree wildly on God's character and what His (or Her) plan for humanity is. Each of these theistic religions is subdivided even further, into countless sects that disagree about what or who God is.

And let's not forget polytheists, who believe that there are many—even countless—gods out there.

For most of my life, I accepted theistic ideas about God: Specifically those associated with Christianity. Christian denominations address the ambiguity of the word *God* by formalizing what they mean when they use it—primarily via creeds or doctrinal statements. There are many conflicting ideas when you look across all these traditions, but a few common notions do arise out of the mess.

The Christian God created the universe, keeps it going, and has a plan for this Creation. God is all-powerful, all-knowing, and everywhere. God inspired the Scriptures. In the Scriptures, God is called Father. Labling God a "father" makes male pronouns are a big deal. But God isn't just our Father—He is also a Trinity of beings. (No one can explain what this means to any reasonable level of satisfaction.)

The Christian God is an active agent in human history. He sent Jesus to die for the sins of humankind, then sent His Holy Spirit to dwell in His followers. God also raised Jesus from the dead, and then Jesus ascended into heaven in front of witnesses. Someday, God will send Jesus to Earth again to judge the living and the dead.

When I lost my faith, these once-unquestioned assumptions began to seem like impossible absurdities. I couldn't just snap my fingers and believe them again—all the challenges I'd learned as an atheist remained.

So I started my search by looking at the most basic Christian claim about God: God created the universe.

This claim is found in the very first verse on the very first page of the Bible. But when I caught my first glimpse of God, it didn't come from Genesis. It came from cosmology.

When we look deep into the night sky with telescopes, we find that all galaxies are moving away from one another and that the rate of their outward movement is increasing. We measure this phenomenon by looking at the redshift of a given galaxy, sort of like figuring out how fast an ambulance is moving away from you by listening to the falling pitch of its siren. With galaxies, we use light waves instead of sound waves, but the outcome is similar. Our universe is expanding, driven by the expansion of space itself.

Now, if you build a mathematical model that accurately describes those galaxies' expansion and run it backward for billions of years, everything begins to converge to a single point: something scientists call the Initial Singularity. The Initial Singularity (the predecessor to the Big Bang) is an unfathomably compressed state in which our entire universe—every galaxy, star, planet, molecule, atom, photon, and quark in every direction as far as we can see with any instrument—fit in a space smaller than a sugar cube.

Physics gets weird when matter, space, and time are compressed to that degree. In the Initial Singularity, the laws of physics didn't exist as we know them now. In fact, back then, space-time was so compressed that matter and energy were the same thing, and the four fundamental forces of physics (gravity, electromagnetism, and the strong and weak nuclear forces) were just one unified field or force. There was no light and no dark, no separation of space, matter, or energy. There was probably no time, either. The clock wasn't ticking, nor was it *not* ticking.

Everything was one, a single-field energy-space thing with the potential to create everything. We don't have language or mathematics to describe this great mystery. We just call it the Singularity.

When I read about the Singularity, I think of God. We're talking about a unified energy that caused everything to be, that is beyond our language and our math, beyond our very imagination. This thought drives me to a state of profound reverence and awe. I was there, billions of years ago, in that Singularity, as were all my ancestors and descendants. Every star that's been born, every star that has died was there, too. So was every particle that makes up every atom in the universe. All was there, together, in the beginning of everything.

God as the Singularity is plausible in physics and similar to the theological idea of God as the Prime Mover or Source of all. God as a Prime Mover was an idea first put forth by Aristotle, but medieval thinkers such as St. Thomas Aquinas adapted it to Christian theology, and Paul Tillich married it to naturalistic philosophy in the middle of the 20th century. Theologians argue that the universe couldn't have come from nothing; therefore, something must have caused the universe to exist, and that something is God. Atheists reply that if it's possible for God to exist without being created, the same could be true for a Singularity or a quantum vacuum or any other state necessary for the Big Bang to occur.

But when I studied cosmology and astrophysics looking for God, it seemed to me that there was, ultimately, no reason to decide between the two. I've heard theologians describe God as the "Great Mystery," and scientists generally agree that whatever predated the Big Bang sits behind an impenetrable veil—the temporal edge of the universe. Either way, we're talking about something that our existing methods of measurement can't describe.

This distinction is mostly technical, though, because as compelling as the concept of God as a Singularity or the Source of all is, it isn't effective for sustaining faith of any kind. That idea might provoke awe and wonder in us, but it doesn't give us a God we can seek in worship or encounter in prayer. And it doesn't explain how God could wash my feet on the shore of the Pacific 13.77 billion years later.

This is why theologians don't just name God as a Source of all. They also claim that God is the "Ground of Being," meaning that God doesn't simply create the universe—he also sustains it to this day.

Of course, physics has its own account of the mechanisms by which we exist now.

Early in the 20th century, Albert Einstein demonstrated that matter and energy are made of the same basic stuff, and that not only is everything that is "solid" in the universe made up of mostly empty space, but that what little actual "mass" there is only exists because some particles interact with a universal, invisible field called the Higgs field. The reason you and I exist is that most of our bodies' particles create some kind of quantum drag against an invisible Higgs field that makes them slow down (from light speed) and gain mass in the process. That's at least as weird as anything in Genesis.

Cosmology describes a force that created us and then transformed itself into a system of forces and energy that continue to sustain the universe. This sounds at least a little like what Paul told the people of Athens: "In him we live and move and have our being."

For the first time since my faith fell away, cosmology gave me an understanding of God that could pass my own skepticism. This was the God of Einstein, a God who can be found in the orderly, elegant mechanics of the cosmos. I was under no illusion that this God was close to anything resembling Christian orthodoxy—in fact, I knew this idea of God was decidedly heretical to many Christians.

I didn't care.

Over three years after losing my faith, I had finally arrived at an answer to the question "How do you know God is real?" I know God is real because I see the *work* of God via telescopes, space probes, and particle accelerators. Instead of fighting science or trying to filter science through my understanding of God, I discovered that you can begin by accepting scientific evidence—and, therefore, scientific accounts of how our universe came to be—and still see the face of God.

So I decided to boil down my understanding of God into a short phrase I could memorize. This phrase could ward off my own skepticism when I prayed, but it would also help me explain myself to the two groups I straddled: Christians, most of whose essential teachings I no longer believed, and skeptics, who were suspicious of my renewed interest in faith. If anyone asked me what I thought about God, I could tell them in simple terms:

God is at least the set of forces that created and sustain the universe.

In philosophy, statements like this are called axioms, ideas that can be accepted without further inquiry.

Even as I wrote it, I knew that my axiom for God was incomplete. Electromagnetism doesn't inspire sacred texts or come to Earth incarnated in a man. There was nothing in my axiom about a Trinity, a begotten-not-made or an incarnated and resurrected Savior. If someone proposed this as the standard idea for God at the Council of Nicaea, he'd probably be excommunicated or executed.

But when you're working with an absence of sufficient evidence, sometimes incomplete explanations are all you can offer. It's not unscientific to admit the limitations of one's knowledge—in fact, the willingness to do so is key if you want the scientific method to work. Here's an example.

Our two most important "rule books" in physics don't agree with each other.

Einstein's theory of relativity does a great job explaining

the counterintuitive behavior of time at very large scales or very high speeds. It seems unbelievable to us that time slows down as you move faster—that when you sit in an airplane seat flying through the atmosphere, the seconds tick more slowly than when you're resting in your favorite chair at home. But it's true, and it's not just theoretical, either. When we make GPS satellites that can orbit Earth at thousands of miles per hour, we have to accommodate the effects of relativity, or else the blue dot on your iPhone would become less accurate with each passing second as the clocks on GPS satellites slide out of sync with clocks here on Earth.

But at very small scales, smaller than an atom, we use a different set of rules: the Standard Model of particle physics. The Standard Model does a fantastic job describing what we observe in the quantum world, and the more data we gather, the more scientists agree that the Standard Model describes reality more accurately than do newer ideas in particle physics.

This is a huge problem, though, because the Standard Model works by pretending that gravity doesn't exist. No, seriously. The only way we can make the math match our observations at the quantum level is to pretend that there's no such thing as gravity. This is kind of like building a defensive strategy in football that assumes there's no such thing as a quarterback. It works great for defending running plays, but what do you do when the other team drops back to pass?

We know the Standard Model is incomplete, but we use it anyway. That's because no more complete, competing model stands up to the tests we perform in particle accelerators. We often think of science as an all-knowing view of reality, but that's now how science works. Science relies on a trial-and-error approach that's dependent on failure—on being wrong but admitting it. Key to its effectiveness, science requires its practitioners to admit what they don't know as readily as they proclaim what they do.

This first axiom was my theological equivalent of the Standard Model. Even if it didn't come close to explaining everything I observed, or everything Christians claim in their theology, it was something I could work with. It let me say, "I believe in God," without crossing my fingers or adding a lot of disclaimers, and there was freedom in that. I could approach God as something other than a convenient self-deception, and I could talk to God in the unself-conscious way I did before I lost my faith.

The phrase "I believe in God" often moved me to tears.

The God of my axiom could explain why there was something instead of nothing and how the universe came to be full of galaxies, stars, and planets. This God could even explain why there was life on Earth.

But I needed to go deeper if I wanted to explain the myriad spiritual experiences humans have every day. Cosmology and particle physics don't explain what happened when I heard Jesus speak to me or why I felt the light that night on the beach. They don't explain the God people encounter when they show up at church on Sunday morning. To find a God who knew me, loved me, and reached out for me—which is to say, the one most people are talking about when they say "God"—I had to shift my gaze from galaxies and gluons to a realm far more intricate and complex.

The human brain.

CHAPTER 10

The God We Can Know

My mom talks about God as she would a dear friend. Whenever she and I are catching up, she frames the conversation by telling me what God has been saying about events in her life, or what God is teaching her. To my mom, God isn't some distant, impersonal force—God is her closest companion.

It's crazy if you stop to think about it. My mom has never seen God. If a grown man were to go around saying he had an imaginary friend, we would think he was delusional, mentally ill. Yet millions of people in America casually mention talking to God, and no one bats an eye.

Some of us experience God as a direct presence in our lives. Others never experience God at all. More and more of us once experienced God but found that those experiences faded as we learned more about the world.

Why?

For as long as humans have believed in some kind of god, someone else has questioned those beliefs. Our earliest religions were animistic, attempts by our brains to understand life-changing events such as why the rains came one season but not the next. Even in those early cultures, I imagine there was some member of the tribe who questioned the shaman's interpretation of what the smoke column was saying about the rainy season. Today, atheism is arguably more widespread than at any point in human history—though it's largely a phenomenon of the Global North, Western civilization.

For a long time, atheism was nothing more than a lack of belief in any god, as opposed to a distinct philosophy or ideology. But in the last 10 years, a more aggressive type of atheism has gained popular acceptance, one that sees religious belief as delusional and contends that human society would be better off without religion. In books such as Richard Dawkins's *The God Delusion* and Sam Harris's *Letter to a Christian Nation*, some of the most brilliant scientists of the modern era declare that instead of improving life, religion actually harms us by undermining our ability to make ethical decisions, accept scientific ideas, and transcend our innate drive toward tribalism. The New Atheists consider faith to be a "virus of the mind."

A lot of people believe this. Even religious people seem to find this argument convincing—convincing enough, at least, to make them self-conscious about their beliefs. Many harbor a secret (or not so secret) fear that they're wrong about God—that their relationship with God is nothing more than a security blanket, or that God is a benevolent imaginary friend who helps them cope with the world.

Some process this doubt internally, and others respond with

defiance, masking their insecurity with over-the-top bluster. You see this in the explosion of resources designed to defend Christianity from rational attack. An entire apologetics industry has appeared, pumping out books, classes, videos, and seminars to teach Christians to confront the claims of atheists.

Faith itself has been profoundly altered by the rise of New Atheism—and research validates this notion. In some studies, when modern believers were asked to explain their faith, brain scans showed that they used similar parts of the brain as they did when lying or trying to sell something. Before sharing about an encounter with God, modern Christians often say things along the lines of, "I know this sounds crazy, but . . ." or, "I know you won't believe this, but . . ." Believers tend to apologize for their beliefs, even as those beliefs continue to define how they see the world.

Are the New Atheists right? Is faith some form of mass delusion? Many skeptics would say that the self-conscious expressions of modern Christians are a sign that faith isn't tenable in a world informed by science. Thankfully, neuroscience offers us a powerful view into what religion is and how religious beliefs form in our brains—all while explaining why some people believe and others don't.

It's difficult for science to study God. Science doesn't speak to the supernatural, and most people's ideas about God are decidedly supernatural. Scientists are generally skeptical of phenomena that don't leave behind physical evidence.

But, given the impact that spirituality has on the well-being of people and society at large, scientists need to study God, or at least the way God influences human behavior. And, thanks to advances in neuroscience and brain imaging in the last few decades, scientists have begun to study God by observing how religious experiences affect the human brain. Some leading neuroscientists involved in this research coined a term for this discipline: neurotheology.

Neurotheology doesn't try to prove or disprove God. It's a pragmatic field entirely devoted to studying the effects religion and spirituality have on human brains. Most of this research involves scanning the brains of people while they pray, meditate, perform religious rituals, or answer questions about God, theology, or the meaning of life.

In research recounted in his book *How God Changes Your Brain,* neuroscientist Andrew Newberg found two basic brain networks that work together to make God real in human minds. First, a network identifies God as an object that exists in the world. Another overlapping network then identifies your relationship with that "God-object."

This relational network is dispersed throughout the brain. The frontal lobe contains all ideas about God—the logical values of faith. The amygdala allows people to fear God or God's wrath, while the stratum and anterior cingulate cortex allow people to feel safe with God and experience God's love. Finally, the thalamus coordinates this entire network and is most responsible for making God appear objectively real.

That this network is complex explains much about our faith. It explains why people with higher activity in their frontal lobes will be drawn to apologetics or theology—they want to know how God works. On the other hand, people with higher activity in their limbic systems will know God through feelings and have little concern with rational justifications for God's existence. They know God because they feel God.

Either way, in most brains, God is not an idea. Instead, God is a set of experiences and feelings attached to an "object" or notion that is closely associated with one's identity. Many people today are dismissive of the whole idea of God and question its relevance to modern society. Some atheists go so far as to deem religion a harmful force in the world. This claim is an overreach that fails to accommodate differences in spiritual beliefs and motivations. It's a rhetorical argument poorly supported by science.

That's because when we look more deeply at the different neurological conceptions of God, we see both the validity of atheists' critique of faith, as well as the real, positive effect faith has on people.

There may be countless religions, sects, denominations, and ideas about God, but as far as your brain is concerned, they all fit within two categories: the Angry God and the Loving God.

The Angry God is the neurological model that forms when you understand God as primarily wrathful or angry—or even just. The Angry God is a god of judgment, who punishes the wicked and smites the wayward. This God demands worship and submission and rewards the faithful in time. In the Bible, this understanding of God is often associated with the Old Testament, and in the landscape of Christian denominations, it finds its clearest expression among modern Fundamentalist sects.

This God is kind of a jerk.

When you experience God as being primarily angry, this experience shows up in your brain. God becomes highly associated with activity in the amygdala. You have more stress, and you anger more easily. It becomes difficult for you to forgive yourself or others, and you become fearful or angry toward those who don't think, look, or act like you.

It's not all bad, however. The Angry God is great for impulse control. If you believe God might punish or smite you, it may motivate you to kick bad habits or develop healthy practices in your life. This is why you so often hear stories of people whose recovery from addiction involved a dramatic conversion to Fundamentalist religion. The Angry God demands change, and this often leads people toward a structured life of service.

Unfortunately, the Angry God is ripe for exploitation. When life's meaning or one's eternal destiny is closed associated with such a God, it's easy for authoritarian systems—whether churches, governments, or even terrorist organizations—to drive people to dangerous behaviors.

The Angry God is exploited by political parties to turn out the vote.

The Angry God tells terrorists to commit acts of mass violence.

The Angry God then tells Christians to fear Muslims.

I believe that many of the New Atheists' critiques of faith come down to the ways in which this ancient, powerful conception of God has often merged with political systems that prize the advancement of an ideology over the good of people. The Angry God plus authoritarianism can be a toxic, dangerous combination that drives religious people to reject science, demonize outsiders, and even commit acts of violence.

Thankfully, the Angry God isn't the only God on the block.

The Loving God is often found in the New Testament of

the Bible, in modern spiritualism, and in Eastern religions. The Loving God is a gracious deity who forgives and nurtures humanity. This God delights in creation and adores humankind.

The Loving God affects the brain in ways that are remarkably different than the Angry God. People who focus on God's love develop thicker, richer gray matter in their prefrontal cortex and anterior cingulate cortex. This development offers them better focus, concentration, compassion, and empathy. They have lower stress levels and lower blood pressure, and it's easier for them to forgive themselves and others. Over time, they even show less activity in the amygdala.

Even more, people who believe that God is loving will eventually develop a characteristic asymmetry in the activity of their thalamus. When that happens, God's love becomes implanted in their sense of identity, and they begin to see the world as being basically safe. This not only allows the believer to experience peace—it also elevates her capacity to take risks for the sake of others. For those who know the Loving God, the risk of being hurt in relationships is less important, because God's love will transcend that hurt.

Most religions involve an understanding of God that includes both love and anger. It's possible to have some combination of these two images of God active in your brain, just as it's possible for a person to transition from one image to the other over time.

I find this neuroscience comforting. First, it helps me understand what causes faith to either go wrong or become a positive force in society. It also tells me that, contrary to some claims, there's no scientific evidence that religion is bad for people. Saying "religion is bad" is a lot like saying "eating is bad." Eating *can* be bad, but it depends on what you eat or how much you eat. Religion *can* be bad, but it depends on how you view God and how attached your faith is to an authoritarian system.

This neurological spectrum between the Loving God and Angry God is especially fascinating in the context of the Bible. Both the Old and New Testaments describe God's wrath and mercy. The authors of Scripture described God in ways that were consistent with their own neurological model, and the resulting narrative explores both perspectives. We'll explore this in much greater detail in a later chapter about the Bible.

Neurotheology shows us the folly of viewing the battle between faith and skepticism as a war of ideas. More than that, it shows us that most critiques of faith tend to be about the effects of authoritarian systems built on an Angry God model. When atheists criticize oppressive religious systems, I stand with them. But to paint all faith with the same brush is to oversimplify the matter, and this view ignores the insights of neuroscientists and anthropologists who find merit in healthy spiritual expression.

Even with those insights, doubt is on the rise in the West. But neurotheology doesn't just tell us how God develops in our brains—it also reveals how God dies. To illustrate this, let's start by looking at religious doubt from a neurological perspective.

Have you ever fallen in love? You know the feeling, when you've just entered a relationship with someone and can't get enough of being around him or her. When you're together, life seems amplified and joyous. When you're apart, this person is always present in your thoughts.

If a neuroscientist were to scan your brain while asking you to think about this special person, the activity levels in different parts of your brain would reveal a lot. Among other things, you'd likely show elevated activity in the parts of your brain responsible for affection or compassion, and also in parts associated with physical attraction. Even when your new partner isn't around, your brain still simulates the feelings that you have while in her or his presence. That's because your brain has created a neurological model of that person.

Over time, the initial mutual attraction between you and that person can grow into something more lasting—a stable, long-term relationship—and your neurological model will grow with it. This is why the best relationships are full of activities that reinforce this bond: walks down the block while holding hands, dreams shared over a meal. These activities deepen our association with the "feeling" parts of our brain, making us feel closer to our partner. Over time, they also help us form a fairly accurate model of that person in our own brains, which helps us anticipate how the other person will think, feel, or behave in different situations. When you pick out a gift for someone that you *know* he will love, you're drawing on this high-fidelity map that has formed in your own neurons and synapses.

Also over time, relationships are often tested by conflict. Maybe you want to vacation in Paris, while your partner lobbies for New York City, or you find that you have different ideas about parenting or retirement. The conflict could even be a minor misunderstanding about schedules or expectations. Whatever form it takes, however, conflict can change the dynamic of a relationship if it's not handled well. Often, unresolved differences can build up so much that they lead to a situation in which two people care deeply for each other, but both wonder what's going wrong with the relationship.

When cherished relationships falter, we often employ

analytical thinking to figure out what went wrong, so we can fix it. We try to figure out what was said or done to bring us to that point. We start thinking about possible solutions or work to understand our own feelings rationally. This kind of thinking happens in the prefrontal cortex.

A few chapters ago, I described the brain as a gigantic company or government operating out of a single immense building. Remember, your prefrontal cortex is the brain's CEO. It views life as a spreadsheet—it's all data points: risk, reward, investment, return. That's great for deciding what bank loan is best, but it's not great at fostering healthy connections with other people.

So, while the impulse to rationally analyze a relationship's problems is a noble one, it often ends up being destructive. We can become obsessed with this fact-finding mission to the point where we neglect the activities that are best at making us feel in love. Instead of enjoying the presence of our partner, we spend all our energy replaying and analyzing every word she or he said.

Over time, this focus on analytical thinking at the expense of affectionate activity reshapes the neurological image we hold of our partner. Our prefrontal cortex does its best to solve the puzzle, but in doing so it weakens the involvement of the more emotional parts of our brain that fire when we think of the one we love. In this way, you can actually recondition your brain to stop experiencing love for your partner. Is it any wonder, then, that people wake up one day and say, "I just don't feel the love anymore"?

In the modern, Western context, we tend to view our beliefs as a set of ideas, which means we often associate mastery of a subject with people who can best articulate the ideas behind their beliefs. This is why, if you ask a Republican why he is a member of that party, he might tell you about the dangers associated with large government, or how low tax rates create jobs and help more people find success. Ask people with an iPhone why they bought it, and they'll probably tell you that Apple's products are more refined and elegant than those of competing brands.

But when you scan the brains of believers, you find that their understanding of God is nonverbal, more akin to a feeling or experience than a set of ideas. This is why Christians are usually stumped if someone asks them, "What is God?" Contrary to what some skeptics say, it's not because these people's belief system is unsophisticated or simplistic. Instead, it's that their experiences with God aren't primarily associated with the language center of the brain.

Trying to describe God is a lot like trying to describe falling in love, and that's a serious problem for people who doubt that God is real. It's also why Christian apologists have such a difficult time reaching those who don't believe. While believers, when asked to focus on God, demonstrate a rich, elaborate neural construct, atheists presented with the same request show nothing but neurological fizzle. The unbelieving brain has no God construct, no neurological model for processing spiritual ideas and experiences in a way that feels real. This is why Bible stories and arguments for God's existence will always sound like nonsense to a skeptic. For the unbeliever, God is truly absent from his or her brain.

Christians tend to view doubt as a precarious spiritual condition, and their prescriptions vary depending on denomination. Many Evangelicals interpret severe, persistent doubt as a "sin problem," best eliminated via repentance. Mainline Christians are more likely to speak of a "dark night of the soul" and prescribe patience or Scripture reading. But neurotheology treats doubt as a neurological condition and would instead encourage people to imagine any God they can accept, and then pray or meditate on that God, in order to reorient the person's neurological image of God back toward the experiential parts of the brain.

I'll talk more about this in the next chapter, but this insight was the most significant turning point in my return to God. I now knew that I had to stop trying to perfect my knowledge of God and instead shift toward activities that would help me cultivate a healthy neurological image of God—secure in the knowledge that this network would help me connect with God and live a peaceful, helpful life.

Anthropologist Tanya Luhrmann goes as far as encouraging people to approach God via "play." She compares the process of identifying a seemingly impossible idea (an unseen, all-powerful being that listens to us) to the story of *The Velveteen Rabbit,* wherein a toy rabbit becomes real via the love of a child who believes it to be more than a toy—in fact, a friend.

That might sound strange to religious folk, but this kind of mental activity is essential in all kinds of idea formation. Highly creative people play around with ideas in their heads all the time: how to decorate their houses, what image to use in an advertisement, or even what precise words might explain something in science. Is it so surprising that play can have a role in connecting with God?

Doubt can't be beaten with willpower. My inbox is full of missives from people who want to believe in God, but their faith is crippled by doubt. If believing in God is important to you, research says that you can start by pretending God is real, giving your brain something to work (or play!) with as you build a new neurological image of God.

In neurotheology, I found a God who, far from being distant, as in the Singularity, was as near me as my own synapses. These insights—how naturally God develops in human brains, and how faith can benefit people—led me to update my axiom for God:

> *God is at least the natural forces that created and sustain the universe as experienced via a psychosocial model in human brains that naturally emerges from innate biases. Even if that is a comprehensive definition for God, the pursuit of this personal, subjective experience can provide meaning, peace, and empathy for others.*

The "Ground of Being/Source of all" from the last chapter created a universe that shows a curious tendency to develop into matter organized into galaxies and star systems with planets, and in the case of at least one planet, things were organized enough for life to spring up and start to evolve.

In time, life developed intelligence and brains complex enough to have subjective experiences. We learned to question why things happened and to try to appease the unseen forces that could decide whether we and our families lived or died. We feared the god of thunder and pleaded to the goddess of rain.

This was a completely new way to process reality, and it appears to have been rewarded by evolution. Some scientists theorize that faith produced social cohesion or obedient children, but for whatever reason, evolution rewarded religious faith.

We also know that when this faith is centered around a loving God, it has positive effects on the believer's emotions and actions. God may be nothing more than a way that human brains interpret reality, but that experience is beneficial. People pursuing God aren't wasting their time.

My axiom is still woefully short of Christian orthodoxy, but its intention is not to justify the Christian faith rationally. It's a way for doubters to practice faith without feeling foolish—because a meaningful understanding of God is built not on rational propositions, but on the "muscle memory" of times when God (or gods) felt real. It's a shield from the most corrosive form of religious doubt: an obsession with analyzing God as a fact.

My axiom is a planted seed, a way for skeptics or doubters to pretend God is real long enough to experience God directly.

But how can we nourish this seed to sprout and grow?

CHAPTER 11

Teach Us, Neuroscience, to Pray

The call came on a morning like any other. I was in my home office, writing this book, with my phone set to Do Not Disturb, all the while ignoring any voice mails. The caller was so desperate to reach me that he called Jenny and asked her if she could track me down.

It was an old friend from our old church—the church I'd lost my faith in. He and I hadn't spoken in years. By now, that church, my folks' divorce, and most of the events in this story were fading in my memory's rearview mirror. But this call wasn't to catch up. It was to let me know that my dad has suffered a massive stroke and was on his way to a hospital near Orlando.

Dad had spoken that morning at a conference for first responders and government officials in charge of emergency management. He was sitting at a table, taking in another presenter's talk, when all of a sudden he dropped his name badge, and his eyes glassed over. Someone asked him if he was OK, and he replied with a thumbs-up, right before slumping over in his chair.

The good news was that this happened while dad was in a room with hundreds of first responders—firemen, police, EMS professionals, and two doctors. The bad news was that everyone who saw him collapse kept using the same awful phrase: "massive stroke."

It took over five hours for Mom and I—the ex-wife and the elder son—to drive from Tallahassee to the hospital in Orlando. I wasn't prepared for what I encountered when I arrived. Doctors had called us during our drive and told us Dad had "weakness" on his left side, but what I saw upon arrival was total numbness and paralysis. The left side of his face was completely slack; the right, etched with pain. Dad tried to project good humor, but he complained of a torturous headache and begged for us to bring pain medication and cold washcloths for his head. (Remember, we're talking about a man who once drove himself to the hospital after being impaled by a tree limb. Dad doesn't complain about pain.)

Hours later, when the brain scans finally arrived, they revealed an extensive right-side stroke, the kind from which the prospects for recovery are slim. I asked, "Doctor, will he ever walk again? Will he be able to move his arm or feed himself?"

The answer: "We don't know, and we won't know for some time."

What else was there to do but pray?

So we prayed, and we asked others to pray. I actually got on my knees—literally kneeling—and asked God to help my Dad.

I did this even though, to this day, I've never resolved Richard Dawkins's milk-jug-experiment dilemma. I did this even though the idea of a God who intervenes when we pray makes me wonder why some prayers get answered with "yes" while others get answered with "your father will never walk again."

Dad's gains were marginal over the first couple of days, but we celebrated and thanked God for each one. Many days later, he stood with assistance. Erratic, wild movement returned to his left arm. And I can remember, as clearly as if I were standing there right now, the first time he said he could feel a hand on his cheek. Tears rolled down my own.

And I thanked God. I thanked God as if the Almighty had appeared in the room and touched Dad with a visible, all-powerful hand. My gratitude had to go somewhere, just as my fear had to go somewhere on that first day, when I saw Dad's mouth droop.

When I came back to faith, prayer returned to me immediately, like an old habit. I prayed all the time. I talked to God the way I would talk to my best friend. But still, these prayers were haunted.

I didn't understand how God could answer prayers at all—or even hear my prayers. Any attempt to build a theological framework for intercessory prayer only served to weaken my tenuous faith. How could God help me but not some other, better person in far more dire need? Why do I get a miracle when someone else doesn't even get clean water?

And that doesn't even touch upon the science. At best, scientific inquiry reveals that prayer has only a modest benefit in helping heal illness. More commonly, studies show no benefit at all, or even a slight detrimental effect on people who know they are being prayed for. These people show a slight, but statistically significant, lower recovery rate compared to that of a control group who was not being prayed for. The reason for this effect is unknown, but researchers hypothesize that it may stem from a "performance anxiety" that raises stress levels and, therefore, inhibits healing.

Still, Americans continue to have a deep faith in the effectiveness of prayer. In one NBC poll, 76 percent of respondents said they believed that prayer has the power to heal—a figure that is backed up by the findings of other polls. A whopping 62 percent said they pray or mediate regularly. Even in an age when religious affiliation is declining, prayer remains popular.

People pray for the things that matter most to them. A LifeWay (a Christian resources group) poll found that among people who pray, 82 percent pray for their family and friends, and 74 percent pray for help with personal problems or difficulties.

They pray about their fantasies and more practical dreams: Two of ten said they'd prayed about winning the lottery, while more than half have prayed for a new job.

They pray despite mixed beliefs on whether or not prayer works: 56 percent of people who pray said that most or all of their prayers were answered, while 37 percent said only some of them were.

But I felt the most affinity for the 14 percent of people who said, "I don't know."

To explore this tension, I asked my Facebook followers to tell me about times when God answered their prayers (or didn't), or when prayer helped them feel connected to God. Some of these stories were heartbreaking or heartwarming. Some were both.

Many of the stories involved healing, either physical or emotional. I heard about fixed knees and backs, cancer remissions, fading burns and warts, and pregnancy crises averted. I also heard about saved marriages and jobs, income appearing when most needed, people finding their soul mate, even one man who walked away from a horrible auto accident—a miracle he attributes to God.

While these stories are touching, a skeptic would note that most can be explained away with relative ease. Maybe the knee injury felt worse than it was, or the cancer was going to go into remission anyway. Maybe the ultrasound dooming the baby-to-be was misread.

For prayers about relationships and financial circumstances—well, those are even easier to explain. Job offers happen. People who care enough to pray about a marriage or relationship will also probably care enough to work to fix it.

But some of the stories weren't so easy to dismiss. Jason told me about a severe allergic rash that faded before his eyes while a pastor prayed for him. Kevin told me about his sister, who as a child had a chronic issue with painful warts. The warts kept coming back after the doctor removed them, so one week at church, she went forward for prayer and asked for healing. Over the next few days, all the warts faded.

In stories like Jason's and Kevin's, there's a short period of time between petition and a specific, measureable response, making it more difficult to dismiss the possibility that prayer was involved in the healing. But each of these answered prayers had a tragic counterpart: a story of someone who didn't find healing, reconciliation, or peace.

Jane told me about being out of work, and how neither she nor her husband could find a way to make ends meet. They were preparing to declare bankruptcy and lose their home. Mark told me about a five-year-old boy with brain cancer, who had 1,500 people praying for him but died, anyway. I heard dozens of stories about strokes, inoperable tumors, freak accidents, and times when someone begged God to preserve the life of a loved one, only to watch that person suffer and die.

When someone prays to God about a serious personal trial, and the prayer goes unanswered, the effect on his faith is often devastating. All of a sudden, God seems distant, uncaring, or even cruel. Stories of other people's answered prayers become unconvincing or, worse, a painful reminder of his loss. Think about how common the admonition to pray is in our culture. For people who've felt neglected by God, prayer becomes the silliest of superstitions, a false source of hope in an uncaring world.

But some people with similar stories arrived at a different perspective. About a third of the tales of loss conveyed that the suffering was profound but also told about the good it created. These folks spoke about how loss and pain at first seemed like unanswered prayers but in time caused such growth in their lives that they came away believing that God works in us, even in the most difficult moments of life. This growth was not just an increased capacity to cope with pain—it was also a greater inclination to appreciate and love others in real, tangible ways. Psychologist Viktor Frankl calls this a "redemptive perspective on suffering," and research has backed up its effectiveness. Somehow, when we see the good that can come from suffering, we suffer less.

One thing I found particularly striking was that when people experienced a prayer going unanswered, or being answered via a redemptive perspective on suffering, the way they prayed changed. Prayer became less about asking God for something and more about being in God's presence. This presence was more than the numinous sensation that God was near. It was also a calming, peaceful outlook that encouraged the petitioners to see the world with loving eyes.

People who had experienced this talked about silent contemplation, meditation, or Christ's supplication on the night of his arrest: "If it's possible, let this cup pass by me." (A request that, by the way, wasn't answered.)

Praying this way might not change your circumstances. But research and experience show that there's a good chance it will change you.

So, here's a review of what we've learned about prayer so far:

1. Studies haven't show prayer to be effective in affecting illness.
2. Still, belief in the power of prayer remains high in the West, even as religious affiliation and church attendance fall. More than half the people who pray believe that most or all of their prayers are answered.
3. This widespread belief that intercessory prayers are answered means that when someone believes a prayer of his has gone unanswered, it often sends him into a crisis of faith.
4. Others, in time, find a redemptive perspective on the suffering associated with an unanswered prayer, which strengthens, but also changes, their faith.

I still use intercessory prayer. I still pray for people in troubling times, like my dad after his stroke, and I still pray for people in good times, that they be blessed and grow. But I no longer do so within the framework of my pre-atheism faith—the framework that casts God as something like an awe-inspiring genie with a divine will, who hands down decisions that are inscrutable from our human perspective.

These days, my intercessory prayers are an act of surrender—a way to voice my hopes and my hopelessness, my power to act and my powerlessness. When I pray for things I hope for, I am searching for ways I can act to make a situation better. When I pray in situations I find hopeless, I'm searching for that redemptive perspective.

And here, research throws a bone to intercessory prayer. Studies have shown that people who pray to God about problems in this way achieve a positive emotional effect, similar to if they'd seen a therapist. Prayer helps us grieve what we've lost, forgive those who've hurt us, and maintain a positive outlook on life. Intercessory prayer is useful and beneficial in helping people surrender to, and process their emotions, and heal from trauma.

But there's yet another type of prayer to consider. Unlike traditional Christian prayer, contemplative prayer is a nonverbal experience. It's the act of being still before God. In contemplative prayer, the goal isn't to petition God to act, but to be in God's presence.

Many of the stories from my Facebook followers illustrate this shift in motivation: People who move from primarily intercessory prayer to primarily contemplative prayer are less driven to find answers to suffering and more apt to redeem it. They stop looking for God on the other side of answered prayer and begin to find God right beside them—in this place, in the pain, in this moment. There's good reason for this. Brain scans show that these prayers bring God near to us—indeed, as near as our own brain tissue.

In the last chapter, we talked about the "God network" that can be found in the brains of believers. We learned that a healthy neurological image of God provides cognitive, emotional, and physical health benefits. These effects are so profound that they can help counteract the effects of stress and distraction on people navigating the modern world.

What I didn't fully explain is how that neurological image of God gets there or how it develops over time. That's because the best means for creating a God object that grows into a God experience in the human brain are activities that are the subject of this chapter: prayer and meditation.

While theologians may ponder the difference between prayer and meditation, for the purposes of our discussion, I'm going to take a neurological vantage point. Neurologically speaking, prayer is a type of meditation, because it produces remarkably similar brain activity and long-term effects.

We can see this specifically in a series of experiments that directed people from different faith traditions to pray while scientists scanned their brains. Among all subjects—Christians, Buddhists, and nonreligious people who meditate—researchers noted increased activity in the frontal lobe, which is responsible for attention and focus. Buddhist monks show increased activity in the occipital (visual) lobe of the cerebral cortex, thanks to the focused visualizations that characterize their practice. Christian nuns, on the other hand, showed increased activity in the language centers of the brain, owing to Christian traditions' emphasis on spoken prayer.

But among these expected results, there was a more surprising piece of data: both the monks and nuns experienced reduced activity in the parietal lobe of the cerebral cortex.

The parietal lobe is the part of our brain that keeps track of our immediate surroundings and sense of physical presence. It gives us our sense of taste and touch, and it creates an ongoing map that's vital to the lower brain when you're trying to escape a source of danger in your environment. But researchers found that religious people with a consistent prayer practice basically shut down their parietal lobe during prayer. This reduced activity can create the sensation that one is leaving this reality and connecting with something greater and less physical. I've experienced this many times, and I was fascinated to learn I'm not the only one.

But the study's main finding was that prayer and meditation are so similar in the brain that we can describe prayer as a type of meditation. And this should be encouraging, because research shows that meditation is one of the best things you can do for your brain—right up there with reading and physical exercise. Neuroscientists have found that people who pray regularly have thicker gray matter in their prefrontal cortex (that's your brain's CEO, responsible for focus and willpower) and their anterior cingulate cortex (the part of your brain responsible for compassion and empathy). The heightened activity in these key parts of the brain also reduces the responsiveness of the amygdala (the part of the brain responsible for fear and anger). You could almost say that consistent meditation makes you a better person—more focused, more compassionate, and less likely to be angry or frightened.

Meditation lowers your blood pressure and helps you feel less stressed. It fosters emotional healing, and it has even been found to help the body cope with disease. These effects are so pronounced, some studies have found meditation to have a therapeutic effect on people suffering from dementia.

In the case of people who meditate on a loving God, the idea of God becomes part of how they process reality—and this has profound effects on their behavior. When you believe God loves you and loves others, it's easier to take risks and to forgive people. It's not enough to simply believe in God, because only prayer and meditation will turn that belief into a neural network that changes your outlook and behavior. Even when the news cycle is depressing or a situation in your life seems hopeless, you can hold on to the knowledge that God is with you and that the overall arc of life will work out for good.

Most remarkable to me is the fact that regular prayer can work for anyone, regardless of their religious background. Even people who self-identify as atheists are likely to report feeling close to God if they pray or meditate consistently for six weeks.

Prayer has always been important to my faith, and now I knew why: Prayer is the most essential practice for cultivating a God network in human brains—even for those who doubt God's existence. In other words, if you want to know God, prayer doesn't come after you've answered every one of your nagging questions. It comes *before.* So I crafted an axiom to make me comfortable spending time in prayer, even on days when I'm not sure God is real:

> *Prayer is at least a form of meditation that encourages the development of healthy brain tissue, lowers stress, and can connect us to God. Even if that is a comprehensive definition*

of prayer, the health and psychological benefits of prayer justify the discipline.

As with my axioms for God, this prayer axiom isn't an attempt to prove Christian orthodoxy. It's a life raft for people who can't get on board with the supernatural claims about God yet still want to be close to God—a state I'm often in. That's why the words *at least* are so critical. I'm not boxing anything in. I'm just stating what I can comfortably support empirically.

The fact is, I've had many overwhelming, beautiful moments in prayer that I can't explain with empirical evidence. I have felt God so close to me, I was sure I'd see a figure standing nearby if I opened my eyes. I've had moments of unfathomable insight about myself, about others, and about actions I should take in my life. And, yes, I have had moments when it seemed as if God was directly answering one of my prayers.

My experience with prayer fills me with hope. A hope that somehow, when we reach out and try to commune with that great mystery we call God—well, somehow God reaches back. In prayer, I can feel the light of some higher, great, more full-of-light realm reaching out, filling me with love, and motivating me to bring as much light into this world as I can.

Somehow, if I admit freely that it could all simply be happening in my head, it helps set my doubts at bay and gives me the space to keep praying—and in so doing, to keep God deeply rooted in my brain.

So, meditation—and, by extension, prayer—is good for you, can make you feel closer to God, and can even change how you see the world at a fundamental level. It can also help us understand one of the most common questions I get when I tell my story: "Why did God speak to you on the beach and not me?"

In many ways, it's remarkable that I became an atheist. Though I have an analytical bent, I've always had remarkable spiritual experiences. When I was seven, I felt God calling to me, telling me I needed to be saved. Years later, I felt as if God was pulling me with a rope around my chest as I listened to missionaries describe their work in Guatemala—a country I visited soon after. I've even felt God leading me to pray for other people many times in my life, with a conviction I couldn't shake.

As a skeptic, I dismissed these feelings as social conditioning and activitiy in my unconscious mind. But after my experience in California, I've felt God's presence in my life even while wrestling with basic ideas about who or what God may be.

Several factors determine how predisposed someone is to spiritual experiences. We find markers for this aptitude in genetic factors (such as the "God gene") as well as psychological ones (such as ratings on the Tellegen Absorption Scale, a measurement of hypnotizability). Physical fitness works in the same way: Some people have the genetic makeup and tendency toward activity that prompt them to early muscularity, while others with, perhaps, a more relaxed disposition have muscles that develop at a slower rate.

But here's the thing: Just as both kinds of people will get stronger with exercise, it's also possible for anyone to increase her or his propensity toward spiritual experiences. Through consistent meditative practice, each of us has the potential to make our brain more spiritual—even to the point of increasing the probability that we will experience something truly mystical.

When Tanya Luhrman, the anthropologist I mentioned earlier, decided to study what happens in the minds of religious people when they experience God, she did so by immersing herself in a local congregation that was part of a network of churches called the Vineyard. What makes her story so interesting is that not only did she interview and study members of the church, but also she participated in their religious rituals herself, in order to experience first-hand what congregants reported. She started attending weekly prayer classes with women in the church and even took part in a prayer retreat. She participated fully in the practices and beliefs of the people she was studying.

Here's the funny thing: During this study, Dr. Luhrman had a mystical experience of her own. After weeks of intensive prayer and meditation exercises, Luhrman woke one morning and saw monks standing around her bed. That experience amazed her and helped her understand the ways that regular prayer changes how people perceive reality to the point that they feel they can see beyond or through reality itself. Tanya Luhrman didn't accept the theological tenets of the Christians she studied or come to their understanding of God. But by simply participating in prayer exercises over time, she primed her brain for the kinds of experiences that are the heartbeat of every major religion.

People often tell me they want to pray or meditate but don't know where to begin. Some tried to meditate but found the process too frustrating; others tried to pray but didn't know what to say. I know how that feels. During and after my journey through atheism, I felt absolutely ridiculous whenever I tried praying to God.

Just remember, the benefits of prayer are available to you

even if you don't believe. The practice is what matters. Plenty of skeptics meditate for the mental and physical health benefits, and if feeling closer to God or confronting doubt is important to you, prayer is going to be more effective than just about anything else you can do. Prayer might not help you solve the mystery of God rationally, but it may help you encounter God.

Here's the good news: It doesn't take much of a time investment to get the benefits of a prayer or meditation routine. The most effective prayer practice is something you do six times a week for 25 to 30 minutes per day. But just as you wouldn't run a marathon without first running a mile, prayer works best if you begin by taking it slowly. Start with five-minute sessions, and build up over time as you become more comfortable.

Here are four exercises you can start with as you build a prayer and meditation practice in your own life. I recommend these because they satisfy two conditions. First, they've been studied and validated by scientific research as being effective. Second, each of them has a long history within the Christian tradition itself.

BASIC PRAYER: TALK TO GOD

Of all the prayers here, this is the easiest and most accessible. Reflect on the idea that God loves you and delights in time spent with you. Then talk to God as you would talk to a close friend. Tell God about your feelings, wishes, desires, and fears. Tell God your dreams and your troubles. You can even ask God to guide you through difficult decisions.

The most important part here is to focus on God's love. Thank God for loving you and for loving your family, your friends, and your community. Then move to reflecting upon your city, then your country, and on to ever more expansive images of God's love. This practice will maximize the benefit you receive from spoken prayers—especially in the anterior cingulate cortex.

BASIC MEDITATION: FOCUS ON COMPASSION

For most people, contemplative meditation is more difficult than spoken prayer. But this lesser accessibility is offset by its effects. For those who can do it, meditation is remarkably effective at developing healthy brain tissue and lowering stress levels.

The goal in compassionate meditation is to relax, become aware, and then focus on a mantra (a calming word or phrase) or an image of peace or compassion. The former could be as simple as "All is well" or "God" or "Love"; the latter could be a field of flowers in bloom or the face of someone you love. Most people find this difficult to do for more than a few moments at a time. As soon as you try to slow your thoughts, a flood of other thoughts comes rushing in. If this happens to you, simply acknowledge that you've had a thought, and gently return your attention to your chosen word, phrase, or image.

If doing this with your eyes closed is too difficult, I've often found that it helps to fix my gaze on a lit candle and imagine that the candle is the warmth of God's love. Choosing a point of visual focus can help quiet the mind by giving you something to occupy your attention.

CENTERING PRAYER: MEDITATION MEETS PRAYER

In centering prayer, you sit in silent contemplation. There is no goal, no insight to receive, just stillness in the presence of God. Start by relaxing and focusing your attention on your breath. Think of each breath as a gift, nourishing you, sustaining you, and requiring no effort to receive. Rest your attention on your breath without trying to control it in any way. If any thoughts or feelings enter your mind, simply notice them and then return your attention to your breath.

That's a basic centering prayer. As your practice deepens, you can move from focusing on your breath to God's love or grace. If you find it difficult to maintain focus in this exercise, scientists have found that adding small, intentional movements or vocalizations can help quiet the mind. (It turns out Catholics and others were on to something with their rosaries and prayer beads!) I often tap my thumb to each finger in turn while repeating, "My God loves me."

The stillness of centering prayer is even more challenging than basic meditation, because it asks you to focus on something more abstract than a word or image. Don't let this difficulty deter you, though. Research shows it can take weeks to get the knack of a centering prayer practice, but once the knack comes, it's one of the surest ways to feel God's presence.

PRAYER WITH SCRIPTURE: *LECTIO DIVINA*

If you're having trouble with meditation but don't know what to say in spoken prayer, the ancient practice of the *Lectio Divina* (or "Divine Reading") can help by offering the benefits of mediation in a more externally structured practice. In *Lectio Divina,* you slowly and carefully pray along with a passage of Scripture. Unlike my manic, four-times-in-one-year Bible binge, *Lectio Divina* is all about a slow, contemplative reading.

Of course, the Bible can be a confusing book for the doubting person. If that's you, don't treat this exercise as a fact-finding mission. Instead, treat the Bible as inspirational literature, and see if its words can bring you insight.

Begin by selecting a Scripture passage. Not anything too long—one "scene" in a narrative book, or one chapter in Psalms will do. A scene in the Gospels where Jesus interacts with someone works well, too.

Next, read the passage three times. On the first pass, listen for any words or phrases that resonate with you. Don't overthink this; it may be something that echoes something that happened recently or a feeling you've had.

On the second pass, read with those words and phrases already in mind. What do you feel? How does the Scripture relate to your life circumstances? Be specific about what in your life resonates with the passage. You can write this down or pray it quietly, but take time to contemplate the words.

On the third reading, think about what action you might take in regard to that situation, as guided by the Scripture. Some ask, "What is God inviting me to do?"

Try all four exercises, and see which ones work for you. Don't dismiss those that may be more difficult. Research suggests that the ones that are harder for you may become more attainable and powerful with practice.

But, no matter which exercises you're doing at any one moment, it's vital to pay attention to your mental outlook. Research tells us that meditation and prayer work best when you prepare your mind by understanding four core ideas.

DON'T JUDGE. One key to a successful meditative practice is to let go of any self-judgment—you literally can't fail at it. Everyone's mind wanders at first, and even with practice, the mind will still wander from time to time. If you fall asleep when meditating, you probably need more sleep, so don't judge your body for telling you what it needs when you create space to reflect. Meditation is not a time to bring along expectations or a performance mentality. Meditation is a time to simply be.

RELAX. The benefits of meditation are associated with states of relaxation. As much as possible, find a quiet, distraction-free place to pray or meditate. Take time to breathe slowly and deeply, and consciously relax your muscles: your neck, your shoulders, your arms, etc. Your cell phone was practically engineered to prevent this from happening, so shut it off or set it to Do Not Disturb. Don't worry: Facebook and Candy Crush will still be there when you get back.

BE AWARE. Prayer is most powerful when it expands our awareness. This is why many types of prayer begin by having you focus on yourself, then your family, then your friends, your community, your city, etc. In meditation, awareness is about noticing what we usually don't:

the pull of gravity on our bodies, the sounds in the room. I hope you're wrapped up in this book right now, but as you read these particular words, you may find that you become aware that you're holding a book or device, and you may become more aware of your surroundings. You can be aware of the fact that you are reading even as you are reading. That's the kind of awareness you strive for in meditative practice.

PRACTICE INTENT. In every exercise, start with an intention—know what you want to achieve. Do you want to feel closer to God? Do you want to forgive someone, be a better friend, or even floss every day? Whatever it is, be intentional, and keep that intent in your awareness throughout the practice. Don't forget the difference between the Angry God and the Loving God here—whatever God you focus on will determine what kind of network you develop in your brain. Be intentional about focusing on God's love and grace.

Honestly, both my Evangelical past and my skeptical tendencies scream at me whenever I spout such new-age-sounding phrases as *practice intent* or *be aware*. But the scientific benefits of meditation are undeniable. Trial after trial validates mediation's power in shaping our emotions, our minds, and even our beliefs and behaviors.

I doubt I'll ever be able to resolve the contradiction between the indifferent universe I understand through science and the intimacy I find with God in prayer. All I know is that even when I was working through existential doubts about God's nature and character, prayer was the one place where God consistently met me.

Perhaps that's why prayer remains so popular, even as religious participation drops. We may feel discouraged by our circumstances, boxed in by our church's expectations, or needled by skeptical scrutiny—but in prayer, we find God no matter where we're starting from.

CHAPTER 12

Jesus

I walked in late, even though I'd budgeted extra hours for the journey. I had stopped for dinner with my friend David, a musician/photographer who'd graciously offered to drive me round-trip between Laguna Beach and Hollywood, but Los Angeles traffic is notoriously unpredictable, and every road between the diner and the studio had turned into a parking lot while we ate.

A year and change had passed since my bewildering reconversion to faith, and I was in that studio to have a talk with comedian Pete Holmes for his podcast *You Made It Weird with Pete Holmes.*

I was an unlikely guest for this program. Pete's usual guests are actors, comedians, and showbiz folk. His one exception to date was Rob Bell, the same Rob Bell from earlier in this story. Rob introduced me to Pete at his house one morning over breakfast by asking me to explain the brain's anatomy, as well as what happens in the brain under a few different scenarios, such as listening to music or eating a taco. When Rob introduced me, he used the moniker "Science Mike."

This label, coined months earlier by my friend Sarah (who you may remember from the conference in Laguna) at a party in Denver, had all but replaced my given name among most of my friends. So, as Pete introduced me to his listeners, it wasn't Mike McHargue in the studio. It was Science Mike—a name now better known to the public than my given one.

Pete and I talked about the brain, cosmology, and the benefits/dangers of religion, but soon enough, we started talking about the central, explosive claim of the Christian faith: the Resurrection of Christ. Pete admitted that it "put a knot in his brain" to think that people believed Jesus underwent a physical death and Resurrection, so I lowered the stakes by saying, "I'm not saying I'm all in on the Resurrection."

It's a sentence that came back to haunt me. At the time, I was still very much a part of the Evangelical church, and you don't have to know much about Evangelicalism to know that it's way out of bounds for an Evangelical to say he's not sure the Resurrection happened. My public admission led to a succession of concerned emails, phone calls, and lunch meetings during which old friends reasoned with me, desperate to secure the eternal security of my soul.

But the truth is, I wasn't even sure about Jesus. Not at all.

When I lost my faith, I was following a collection of scholars, the mythicists, who believe there's not enough evidence to establish Jesus of Nazareth as an actual historical figure. Their analysis is widely quoted in the online spaces where atheists congregate, and this gave me the impression that the mythicists represented a consensus among secular historians—a belief I still held when I taped *You Made It Weird.*

I was a Christian who believed he heard the voice of Christ but was unsure that Jesus was an actual historical figure. That's a big problem if you want to call yourself a Christian.

The original followers of Jesus preferred to be called brethren or disciples, in keeping with the traditions of the Gospels. They'd also use the name "Followers of the Way," as in "I am the way, the truth, and the life . . ." Some scholars believe that the word *Christian* originated as an insult, as a pun meaning either something like "Goody Two-shoes" or maybe "little Christs." But what began as insult eventually became a badge of honor. Christians are all about Jesus of Nazareth; their entire faith is based on his birth, life, death, and Resurrection.

For Christians, Jesus is a big deal. Jesus is the biggest possible deal. Christians are people who follow not just God or holy Scriptures, but Jesus Christ specifically. Even that name, Jesus Christ, starts laying out some major assumptions about the person in question. Jesus of Nazareth was a man. Christ was not. When we say "Jesus Christ," we're invoking two distinct ideas: a person, named Jesus, and the Christ, the reconciling spirit of the Holy Trinity.

To talk about Jesus as Christ is to invoke a whole set of other dependent beliefs, such as sin, salvation, incarnation, atonement, heaven, and hell. The Christian faith is often presented as a test you pass by accepting all these ideas, no matter how much your mind rebels against them, but for someone like me, who has spent serious time pondering the absurdity of a man who rose from the dead, this gets really messy, really fast.

I've heard many a preacher say that you have to take a leap and believe before any of it makes sense. The problem is, I did that. I leapt, I believed, and it made sense. Until it didn't. It stopped making sense when I started studying secular accounts of the life and death of Jesus of Nazareth.

Of course, you know that's not the end of my story. I'm not writing a book against Jesus. I believe again. But I also know that my beliefs about God—and, yes, even my beliefs about Jesus—are far outside what would normally be called Christian orthodoxy.

I have faith in Jesus again, but it is nothing like the faith of my youth.

Christians talk a lot about the importance of "orthodoxy" and "orthodox Christianity," but there are many different opinions about what "orthodox Christianity" is in the first place. The earliest followers of Jesus had rigorous debates about what his life, death, and Resurrection meant for humanity. Today, Eastern Orthodox, Roman Catholic, American Evangelical, and American mainline denominations all hold to different orthodoxies, and many of them have, at some point, considered one another heretics.

But, for the sake of identifying what most Christians mean when they use the word *orthodox,* let's describe a basic way of thinking and believing that most Christians in America would agree upon.

God is a Trinity of beings with a single essence: God the Father, God the Son (the Christ), and God the Holy Spirit. God is perfect and holy and cannot be in the presence of sin. God created the universe, including Earth and all life on it. This initial state of creation was perfect and holy, but humanity chose to break God's laws, and sin entered the world.

God and humanity were therefore separated. God's holy justice requires punishment for sin. In order to reconcile humankind back to God, God sent the Law in the form of the Torah, which included a system of animal sacrifices to cover the sins of humanity. This system formed a covenant between God and humanity—but humanity never kept its side of the deal.

God sent the Christ to inhabit Jesus, the Son of God, to lead him to live a perfect, sinless life as a human. Jesus faced all the temptations we face in life, but his perfect character allowed him to resist. In his life and teachings, Jesus completely fulfilled the Law of the Old Covenant.

His rising popularity as a teacher made him a threat to both religious and political leaders in first-century Israel. He was tried for treason and executed on a cross—punishment for sins he never committed. At that moment, Jesus took on all the sins of humanity and reconciled mankind to God in a New Covenant.

In the New Covenant, those who follow Jesus as their Lord receive forgiveness for their sins and at the end of their physical life will go to spend eternity with God. Those who don't follow Jesus don't receive forgiveness and are, therefore, unable to be with God when they die. They go to hell, the only place in all creation where God is not.

Finally, God's Holy Spirit lives in all of Jesus' followers across history, guiding them in an ongoing battle against God's adversaries: Satan and his legions of fallen angels, now called demons.

This orthodoxy begins with Christ as part of the Trinity, and its story features Jesus prominently as the protagonist. Without Jesus, all of humanity is destined for hell, because of the fall of man. You can distill the whole thing down to two words: *Jesus saves.*

The idea that Jesus' death was a sacrifice to erase the sins of humanity is called penal substitutionary atonement. It's the idea that Jesus' perfect, sinless life made him a worthy surrogate to take on the punishment that the rest of humanity had earned. And, indeed, this paints Jesus in an extraordinary light: suffering a brutal death via execution for the sake of others.

Depending on where you locate yourself among the branches of Christianity's ever-growing tree of denominations, this narrative of salvation and the afterlife will take a slightly different form. Evangelicals emphasize salvation as a specific event, while mainline denominations speak of salvation as more of an ongoing process. Catholics believe in a Purgatory where some find their way into heaven after this life, and a growing number of Christians don't believe in an eternal hell at all. Even in light of these varying positions, I think the sketch above is a fair representation of middle-of-the-road American Christianity.

But in this relatively tame picture of God, Jesus, salvation, and the afterlife, there are still problems—problems troubling enough that, when I discovered them, I became an atheist. I've heard skeptics satirically summarize the substitutionary atonement approach to Christianity this way: "God sent *himself* in the form of his own son to sacrifice *himself* to *himself* so that he could save humanity from *himself*." I'll be the first to admit that this glosses over some of the finer points of Christian theology. Jesus, for example, is said to have offered himself willingly as a sacrifice. Many Christians would say that God doesn't send anyone to hell but, instead, that people choose hell by rejecting God's offer of salvation.

Still, you'd have to forgive a skeptic for suggesting that the God of this picture sounds pretty monstrous. This God creates people with the potential to sin, a tendency that is inevitable for anyone born after the fall of man. These humans are born into the world completely ignorant of God—they're blank slates later written upon and shaped by their families and culture. Yet, if a person commits even the most insignificant of sins at any point in his life—say, the occasional lie or stealing some candy in childhood—and does not atone for them, that person could spend eternity in hell.

Imagine if a secular government on Earth sentenced its citizens to death for such tiny infractions. Without question, we would consider it a humanitarian crisis. Yet God, who is said to be the most just, loving entity possible, does worse than execution. God offers eternal torment.

Sure, some sins may justify extreme punishment: genocide, torture, and murder, among others. But the worst actions committed by most people don't even register on the scale of ones that would merit eternal, conscious punishment. Worse still, many Christians go so far as to say that some people are God's "elect," a group called to salvation, while the rest of humanity is sentenced to hell with absolutely no chance for redemption—all because they don't have the capacity to choose to accept substitutionary atonement.

If a car company made vehicles that were destined to crash regardless of the driver's steering, we would blame the company when those cars crashed—not the vehicle or the driver. But penal substitutionary atonement makes humans liable for God's creative action.

The obvious response to these objections is that God offers

a means to escape hell in the sacrifice Jesus made on the cross. Some Christians even go so far as to say that Christ's sacrifice provided universal coverage—that every person who ever lived is able to avoid eternity in hell, even if they're a non-Christian, because Jesus really paid for it all: every sin, by every person, ever.

But this still paints a troubling picture of God.

Let's say that you and I are close friends, but after an argument one night, you stole my car and drove it into a lake. This is a serious crime with a serious penalty—let's say $10,000 in damages and three years spent in prison. Now, imagine you came to me and apologized, expressing sincere regret and grief over your actions. What if I responded by telling you I could forgive you, but only if my daughter took your place in prison and paid the fine on your behalf, because I am a merciful and just friend. My mercy compels me to forgive you, but my justice demands that the crime be punished.

This is the exact picture that most Christians paint of God: a God who offers no choice but to demand punishment for sins. But if a good friend of mine wrecked my car, I could simply forgive that friend without anyone being punished. I'm a nice guy but certainly not the embodiment of perfect love—so why can I forgive with no strings attached, but God can't?

On this question, there's actually some good news. Penal substitutionary atonement may be the dominant Christian view on salvation and atonement today, but the Church has held a number of other ideas during its history. The Eastern Orthodox traditions view sin less as a crime to be punished and more as a sickness to be healed—a view I find compelling. Earlier Church ideas on salvation included such views as *Christus Victor*, which held that Christ's death and Resurrection signaled God's triumph over evil, and even "ransom atonement," in which Jesus' death paid a ransom to liberate humankind from Satan.

These earlier views on atonement are less troubling, for sure, but they still make some really strange assumptions about God—and posit Satan as a major player in how the universe unfolds.

So, I have to ask: What if the reason for Jesus' death wasn't that God's justice demands human sacrifice? What if God wasn't the one looking for blood?

The Old Testament is full of retributive justice—but let's remember, "an eye for an eye" meant to *limit* violence, not encourage it. Prior to laws such as this, vengeance was the normative function of justice, and it often worked out as "your whole village for an eye." Humans have often demonstrated a remarkably brutal thirst for violence, and in many cases, the Old Testament's seemingly harsh vision of justice was a vast improvement over what had come before.

The cross was not God's invention—it was ours. The cross was an instrument of torture, a method of intimidation created by an empire that needed to keep its conquered cities in check. In all our need for an eye for an eye, I have to wonder sometimes if Jesus' sacrifice on the cross is an answer not to God's wrath, but to ours. I have to wonder if God, having listened to us cry for blood, decided to offer his own.

Perhaps Jesus hung on a cross to demonstrate the inevitable outcome of retributive justice in the face of an empire that used violence to expand, that survived only by placing societies under its oppressive heel. Jesus didn't hold up a sword in response to a sword. He took the sword into his side, and in doing so, revealed our brutality for what it was.

Even this view of Jesus' death probably anthropomorphizes God too much. Like penal substitutionary atonement or *Christus Victor*, it places God in a human perspective—standing on Earth, experiencing emotions, trapped inside Isaac Newton's cause-and-effect–based laws of motion and following an arrow of time that marches inexorably from past to present and on toward the future. But physics doesn't support this view of reality. Though Christian orthodoxy may talk about an "unchanging God," a Christ who responds to our sin with an equal and opposite sacrifice reflects a very human conception of time and, therefore, God.

As humans, we experience space-time as a series of moments, with some in the past, some in the future, and one special, fleeting moment called now. But for God to be God, God's reference frame must include the strange, shifting perspective of Einstein's relativity. God's perspective has to encompass space-time not as a series of moments, as we see them, but as an interconnected set of coordinates that always exists.

But when I think about God in this light, I am faced with something so incomprehensible that such words as *being, consciousness,* and *free will* become nonsensical. Even discounting more mind-bending possibilities, such as the extra-spatial dimensions of string theory or the additional "universes" described in different multiverse theories, God's perspective is so foreign to me that I can barely call it a perspective at all.

But this is incredibly comforting. Because in contrast to God's mystery, Jesus makes sense. The Christian idea of incarnation—God in the form of a human—makes this "foreign" God approachable. Jesus gives God a face, a language, and stories we can understand. If there's one thing I've come to believe about the hand-wringing over atonement theory, it's that it misses the value of Jesus Christ. The value of Jesus Christ is in our admirable but flawed attempts to understand divinity from a reference frame we can approach. It's in a God we can know, a God who walks with us, talks with us, and can empathize with our pain.

Contrary to the claim of mythicists, the consensus view among historians today is that Jesus was a real person. Jesus really walked around first-century Israel under Roman rule. He studied the Torah and taught it to others. He very likely studied under another rabbi. Jesus was a dark-skinned man who wandered from village to village sharing insights on how to apply the Torah to daily life. He ate, drank, laughed, and swam. He danced, burped, and emptied his bladder. Jesus of Nazareth was a man. It's a claim accepted in institutions of higher learning. Richard Dawkins himself has come under fire from historians for asserting otherwise.

Sadly, this is little comfort to those of us who doubt. Here's why: Christianity isn't built solely on the life of Christ; it's built on his death and Resurrection. Paul himself said the Christian faith is foolish if the Resurrection didn't really happen. Jesus's emergence from a tomb three days after his death is Exhibit A in the argument that Jesus of Nazareth was also Jesus Christ, who takes away the sins of the world. There's also Jesus' virgin birth, a sign of both his sinlessness and his divinity.

It's not difficult to claim that someone existed. People exist all the time—we've literally got billions of people who exist right now as evidence of that idea. But historians and biologists are pretty clear on the fact that people don't die and rise from the grave. That's an incredibly bold claim, one that would require extraordinary evidence to validate.

The accounts of the Gospels won't do it—they were written too late, and their authorship is too much in question. Paul claims to be both a secondhand witness (to know people who knew Jesus firsthand) and one who had his own personal, mystical experience with Jesus on the road to Damascus (Paul says Jesus appeared to him).

I know people who say they know people who had their kidneys cut out after a kidnapping, but I would be crazy to bet my life on that being true. Secondhand testimony is evidence, but it's not sufficient for claims of this magnitude.

Likewise, if we're limiting ourselves to what can be proved via empirical evidence, we can't treat Paul's claim of a personal encounter with Jesus any differently than we treat Joseph Smith's claim that he could see invisible gold tablets with the Book of Mormon on them. To accept one person's claim of revelation without evidence is to accept them all.

Don't misread me. I'm not discounting the value of the Bible, or even whether it's right to use the Bible's books as evidence. What I am saying is that the evidence isn't strong enough to verify a claim as extraordinary as the one that Jesus rose from the grave.

This is the challenge I faced when I was returning to faith. Finding an understanding of God that was plausible in the face of skepticism was relatively easy, but doing the same for Jesus Christ was not. Jesus, a man who walked the earth in the first century dispensing radical teachings about God? Sure. But Jesus Christ, the reconciling member of God's Trinity? No.

The voice I heard speak to me on the beach was as ethereal as the voice Paul heard on the road to Damascus. It was a powerful voice, one that changed my life. But did it really belong to Jesus? And how could I know?

The only way I could place myself in the Christian tradition was to come to an understanding of Jesus. So, a few weeks after writing my God axiom, I sat down with a pencil and paper and came up with a list of claims I could confidently make about Jesus:

> Jesus was a real person (or maybe a few people) in the first century.
>
> His teachings moved people enough that he began to attract a significant following.
>
> What Jesus taught put him at odds with political leaders, to the point that he was probably crucified. (Crucifixion is not an extraordinary claim; many people were crucified by the Roman Empire.)
>
> The teachings of Jesus inspired a persistent social/religious movement after his death.
>
> Over time, this movement became the largest religion in human history.

Most Christians wouldn't be comfortable describing Jesus this way. But for me, it was a big step up from believing that Jesus was a myth. Accepting Jesus as an actual historical figure whose teachings created the largest spiritual movement in history gave me a relatively reasonable starting point for my decision to follow him.

With this in mind, I wrote an axiom to set a "floor" for what I accepted about Jesus:

> *Jesus is at least a man so connected to God that he was called the Son of God, and the largest religious movement in human history is centered around his teachings. Even if this is all Jesus is, following his teachings can promote peace, empathy, and genuine morality.*

There's no claim here of "begotten, not made," born-of-a-virgin, or even a Resurrection. But what my axiom does say is that, putting aside the question of Jesus' divinity, Jesus was so compelling in how he described God that people attributed divinity to him—and a person who provokes that response is a person who's probably worth paying attention to.

You can be skeptical about the Resurrection and still have an encounter with Jesus that's life changing. Experiments verify this: In brain scans, many Christians show the characteristic brain activity of people who view the world as basically safe. Jesus is compelling because he produced this psychic shift away from superstition and fear in people millennia before we understood what that meant for the brain.

Jesus lived and taught in an era when might made right. The Roman Empire assimilated other cultures via a combination of military brutality and an odd sort of religious tolerance—you could have your gods, but only if they were as subject to Caesar's reign as you were. The Jewish people under this rule responded with either peaceful submission or violent rebellion. But Jesus flipped both scripts.

In the face of violent rebellion, Jesus offered a sermon that said "blessed are those" who were the least powerful people in his society. Instead of peaceful submission, he "turned the other cheek" and "walked two miles when asked to walk one," in order to reveal the cruelty of the Roman Empire and a better way of reacting when threatened. In both cases, Jesus responded to an unsafe world by offering compassion—the kind rooted in the anterior cingulate cortex.

I don't know for sure whether Jesus rose from the dead. What I do know is that there are things I've done that I have a hard time forgiving myself for, but I'm able to do so when I believe Jesus already forgave me. I know that the story of a man who told people to "turn the other cheek" inspires me to approach people differently and that this inspiration affects those who interact with me.

From this perspective, I didn't necessarily invite Jesus "into my heart," as the saying goes. Instead, Jesus lives in my anterior cingulate cortex, the seat of compassion. He reduces my tendency toward selfishness, anger, and fear. He teaches me a more patient approach to life—as he's been teaching those who follow him for over 2,000 years.

Understanding Jesus in this way also helps us approach other assumptions about the ramifications of his life—the ones we talked about earlier in this chapter. Let's start with the idea of sin.

Many skeptics say sin is an inherently flawed idea, but I'm with the Baptists here: Sin is a useful idea in understanding human behavior. I didn't have to teach my children to do wrong. My kids, lovely as they are, figured out how to lie on their own. Even if Adam and Eve never existed—even if original sin didn't enter the world through them—we sin, regardless. Inside each of us is a minefield of self-serving impulses that often cause us to harm ourselves and others.

I think science reveals the truth behind sin:

> *Sin is at least volitional action or inaction that violates human consent or produces human suffering. Sin comes from the divergent impulses between our lower and higher brain functions and is accelerated by our evolution-driven tendency to do things that serve ourselves and our tribe. Even if this is all sin is, it is destructive and threatens human flourishing.*

This is true of both individuals and societies. A person can commit the intentional sin of acting out of prejudice, but a society can amplify that racism (or sexism or homophobia, etc.) into a set of laws and cultural biases that favor one group of people at the expense of another. Sin ravages families via domestic violence, and it destroys societies through geopolitical conflict or civil war. In this light, Christianity's dire warnings about sin's consequences and the necessity for salvation aren't absurd. The issues Christians preached against in the first century—imperialism, violence, racism—are the ones we're still working on today.

So you'll have to forgive me if I cast a skeptical eye at anyone who says, "Humanity doesn't need saving." We are in desperate need of salvation. Climate change, persistent racism, and nuclear war are existential threats as big as anything imagined in biblical times.

I'd be lying if I said my axioms covered everything I believe about Jesus. The voice of Jesus in my ear makes me hope for more than just a man who promoted revolutionary ideas. I hope for an incarnate Christ, a virgin birth, and an empty tomb. I hope for a universe with a moral arc that bends toward justice and not simply chaos. It's not that I would make these other beliefs claims of fact—it's more that my experiences in faith inspire a kind of reckless anticipation.

There's nothing in my axioms about the Resurrection of Jesus, either. But that doesn't mean I still believe what I said when I told Pete Holmes, "I'm not all in on the Resurrection." That's because I've come to realize that resurrection is everywhere I look.

In the beginning, there was a rapid expansion of a Singularity. Around 380,000 years later, there was light. There was also hydrogen and helium and four stable, fundamental forces of physics. Atoms and those forces worked together to birth the first stars from massive clouds of gas, and those stars lived for hundreds of millions of years before they died in explosions that spread their matter across the sky in clouds of gas and dust—now with heavier elements than what existed before.

The forces of physics worked together once again to craft new stars now tightly packed into the first galaxies.

As the cycle repeated, heavier elements formed planets orbiting those stars, emerging from disks of gas and dust like dust bunnies under your bed. In our universe, planets can only exist because a few generations of stars died and were reborn. The rebirth of stellar matter into planets is how our Earth came to be.

This planet, our home, is covered with a film of life unlike any we've yet seen anywhere else in the universe. As far as we know today, it is unique. A blue marble floating in the dark.

Earth's life is fed by a process in which carbon from the air and minerals in the soil are attached together by the energy of photons via photosynthesis in plants. In this process, everything on this planet lives by the constant sacrifice of the nearest star. Every blade of glass, every tree, every bush, every microscopic alga on this planet is a resurrected form of the Sun's energy.

I capture that energy by consuming other things that have died. Every time I eat a meal, the dead matter that made those plants and animals literally gives life to my body through digestion and my metabolism.

One day, I will die, and in time my atoms will go back to giving life to something else. Much farther along the arrow of time, our own Sun will explode and spread its essence across the sky. Our Sun's dust will meet with other stars' remnants and form new stars and planets of their own. The universe itself exists in an eternal pattern of life, death, and resurrection.

It seems poetically appropriate that the Source of all would have left this divine signature on the fabric of reality. In Jesus, I hope for more than just a God with a face or a uniquely gifted moral teacher. I hope for a resurrection that will one day reach every corner of our universe.

As I've said, Jesus is a pretty big deal.

CHAPTER 13

Take Me to Church

It hurts when a good thing goes bad, doesn't it? I mean, it's one thing when your enemy attacks you—that's expected. But when a cherished relationship turns sour, it cuts us to the core and bypasses all our defenses.

I've been through breakups, and I've had friendships wither away despite my best efforts to fix them. But the most painful loss in my life to date was the loss of my beloved Southern Baptist church.

Most of the people at my church only discovered I'd lost my faith after I'd already regained it. Worse, they learned of this change in perhaps the least happy way possible: a series of blog and Facebook posts in which I challenged our prevailing dogma about LGBTQ relationships, the Creation account in Genesis, and, ultimately, the authority of the Bible. Only then did I admit that I had been an atheist for the previous two years and was now in the process of rebuilding my faith.

This caused a lot of chaos—especially since I was dropping

these bombshells while serving as a leader in a Southern Baptist congregation. Some of the other deacons called for my resignation, and some church members wanted me removed as a Sunday-school teacher. These folks were counterbalanced by members who wanted LGBTQ people to feel welcome in our congregation, or who wanted to make room for spiritual doubters to work out their beliefs inside the church.

I'd become a point of contention and division in our church—something I'd never intended but now couldn't avoid. I no longer had the energy to keep putting on a mask and faking my conformity to my church's norms and values. Piece by piece, I was reconciling the person I presented to the world with the person I really was, and I knew the process was healthy and worth the pain for me. But there was no way to deny its collateral damage. My church was hurting, and the more people on the Internet who heard and responded to my honesty, the more pain I saw in my local church.

That's why I called my friend Andy. For years, Andy and I had worked together in the church, teaching Sunday school and playing in the worship band. I was lost, torn between how much I loved all the people in our church and how strongly I felt for all the people I knew were hiding their doubts in pews or, worse, slipping silently out the back door. I told Andy I was tired of being a point of controversy. All I wanted to do was help people and figure out how to follow God.

It was a cold December night, and I was pacing around in my garage, talking on my cell phone and occasionally rubbing my hands together to keep the numbness at bay. I remember the tears in Andy's voice as he told me how he felt.

"It's been hard, Mike," he said. "You love those students and they love you, and you love being a deacon, and we've loved having you, but you're right: There's controversy. We spend more time talking about you than the work of the church. I'm so excited to see where God is leading you, but I'm scared, too.

"I look at your blog and all the comments you get. It doesn't matter that you say a lot that I disagree with; God is using you to do something. Maybe being a deacon and teaching church kids is a distraction now. Maybe God is telling you that your real work is online, with people who can't believe in Him."

"Andy, do you think I should step down?"

"Only you can say that. Let me ask you this: How would you feel if you did?"

I stopped and imagined what it would feel like, turning in a letter that relinquished those things I loved doing every Sunday.

I was surprised at how it felt.

"I'd feel relieved, Andy. I am very tired, and I think I would feel more relieved than anything."

Andy paused for a second, and then he said, "I think you have your answer, Mike."

We prayed and cried together over the phone.

So I sat down with my pastor and told him I wanted to resign as a deacon and Sunday-school teacher. I told him I wanted to do it in a way that didn't make our church look bad; I didn't want anyone to think I'd been kicked out. I can't overstate how kind and helpful the pastor was. He listened, and he encouraged me, even as he made it clear that he disagreed with my understanding of the Christian faith.

There were no torches or pitchforks chasing me out of our

church's leadership. There were just the sad, confused faces of friends who'd been at my wedding or cheered as my daughter's head emerged from the surface of the water in the baptismal font.

I expected everything to get better after I stepped down from my leadership position. Now that the church could distance itself from my opinions, the controversy would be over, and I could go back to sitting among people I loved every Sunday. I thought this would be an end to anonymous letters showing up at my office, or to whispers that followed me down the hall on my way to Sunday school.

But church was never the same again. Jenny and I looked for an adult Sunday-school class, and I was shocked when someone called me at the office Monday morning to tell me the group wasn't comfortable with my presence. We never knew which Sunday at our own church could produce someone trying to "correct" me, and no way could we miss how many shoulders had turned cold. My family felt increasingly alienated in what had once been our spiritual home.

So we did what anyone would do. We left our church. Over the next few months, we learned the near-sacred pleasure of Sunday brunch and the "miracle" of a weekend that suddenly had two days. For the first time in my life, I didn't have a church.

But it broke my heart. Honestly, it hurt worse than my parents' divorce when we left that Baptist church behind. I'd proposed to my wife in that sanctuary and watched her walk down its center aisle on the day we wed. Both my girls were born while we were there, and both were dedicated to God in its Sunday services, under the loving gaze of its pastor. We'd grieved there, been comforted there, fed the hungry there, and built our lives in its community.

To this day, I can't drive by that church without saying a prayer for its every member, wishing them well, and asking God to bless them.

We've all heard stories about people who have been hurt by church officiants committing untoward acts. Those experiences are genuine and important, but my tale isn't one of them. There was no villain here. The people at my church loved me, and the tension I experienced there wasn't caused by a loss of love, but by disagreement. I was openly LGBTQ-affirming, and my church saw same-sex relationships as a rebellion against God. I was an evolutionist, and some of our leaders saw that as an affront to God's authority. There was no way of denying it: Nothing about my theological views fit within the Baptist faith.

My story of breaking up with my church is one in which both sides had the best of intentions. I was following a path toward who I believed God wanted me to be, and my church was being faithful in a way its members understood. But to stop fitting in at a church one loves can prove incredibly painful.

In these instances, the temptation might be to stay and tough out the discomfort. This seems a sensible instinct; after all, any relationship will go through periods of conflict and tension. People grow, change, and disagree. It's unrealistic, even unhealthy, to desire some kind of conflict-free utopia from our faith communities.

But I knew it was time to go when I died a little inside every Sunday. I knew the time had come to move on when others were being hurt by the simple act of me being who I was. I had worried that my church might suffer without me, but I was wrong.

My church suffered *because* of my presence, and it could only start to heal once I left.

Sometimes "church wounds" come from something more sinister than the kind of gradual drift and alienation I experienced. Some churches have a strong authoritarian streak and see their primary mission as defending God. These institutions often act in ways that are emotionally abusive, causing pain and suffering to those in their midst who are shunned, rebuked, marginalized, or forgotten.

My email inbox bears witness to this kind of pain. Kimmy told me she was ostracized for pursuing the "godless" activity of joining a marching band in college. Andrea was driven off the worship team at her church because someone sent in an anonymous letter saying she was "too promiscuous," even though she was a virgin. Deborah was fired from her custodial job at a Christian school because she's gay (the implication being that lesbians are likely to molest children).

I have emails from women who were told they couldn't teach or sing in their church because of their gender. I've read accounts from gay teens who were driven away after coming out. Several said they were ejected from the church after confronting an abusive pastor, or that they suffered anxiety and even PTSD (post-traumatic stress disorder) after being banished from their spiritual communities.

One of the most heartbreaking stories came from my friend Kevin. Here's what he told me:

> When I was 10 years old, my Dad was a music minister and youth pastor at a small church in the Bible Belt. My

Mother worked alongside him in the youth program. I didn't know it at the time, but we were well below the poverty line, and while going through the process of buying a house (using a special military refinance process), we found out that several repairs had to be made before we could live in it. We couldn't pay anyone to make the repairs, so we did them ourselves. One weekend, we learned that a major inspection had been scheduled for the following Monday, so my parents stayed home from church to finish painting the outside of the house.

When the church found out, they were furious that we had broken a commandment and not kept the Sabbath holy. They decided to hold a "stoning service," in which they excommunicated my family along with another woman who was guilty of working at a convenience store that sold alcohol and cigarettes.

How do you explain to a nine-year-old what a "stoning service" is when he's afraid someone is about to literally throw rocks at him? How do you explain that people who were supposed to be friends and family were saying vile things about you? How do you explain why the place your family worked so hard to serve has now made your parents cry? I'll never forget the look on my parents' faces or especially that poor lady whose only sin was living in a small town and taking the only job she could find. I pray to this day that she found a group of Christians who loved her through this, but I suspect that didn't happen.

I know I'm supposed to forgive, but at 37 years old, I'm still angry. It hurts to this day every time I relive that moment in my brain. However, I won't let myself forget. I can't.

Is it any wonder that fewer people go to church every Sunday? Stories like these put flesh on statistical bones, put a face on an abstract issue. The church, the spiritual body of Christ, meant to mend the wounds of those who suffer, instead can become the bully, an advocate for the powerful, and an enemy of the oppressed. A holy commission to go out and make disciples is twisted into a call to make others into enemies in a culture war.

That kind of church is dying, and it needs to. It has little to do with Christ.

And this perversion of the Gospel is especially toxic when we consider its neurological implications. We have learned that spirituality involves deep wiring in our brains. As spirituality organizes itself into religions, it has tremendous power to create close communities, and this power carries remarkable benefits. One study found that people who get involved in a church have a similar shift in happiness as people who move from the bottom quartile of income to the top quarter.

But this power can come at a cost. People who suffer at the hands of and subsequently leave their religious communities are more prone to depression and anxiety, show an elevated risk for suicide, and can even suffer PTSD.

I hope that talking about the way a church can inadvertently harm people doesn't sound like deep resentment on my part. The reason I focus so much on recognizing such wounds and the need to properly grieve them is because I believe that, after prayer, church is the most essential component of Christian faith. But for those who've felt ostracized by a church, the prospect of entering another spiritual community can be terrifying.

When we look at the human brain, we can see how a church would be well suited to nurture people's spiritual growth. A church represents an embodiment of two primal human needs:

an urgency to belong to a community and a desire to experience God. This synergy is powerful, but this power must be handled with great care. What can we do to ensure that it is?

Despite my newfound Sunday-morning ease, leaving my church drove me into a serious funk. The melancholy anxiety I had begun to feel every Sunday at church soon ripened into depression. I had no minister to turn to, so I started going to weekly therapy.

On a cognitive level, I understood the sociological dimension of my faith transition. I was the one whose beliefs had changed, and that change was a violation of an unspoken social code in my church community. My closest friends felt betrayed, and strained or broken relationships became inevitable. But even though I understood why everything had happened, it didn't bring me any peace.

That's what sent me to therapy. Most of the time, if I understand a problem, my feelings fix themselves in short order. But this time, I kept crying at random moments or overreacting to everyday events. My temper was far easier to trigger than before, and my generally sunny disposition could slip into darker places with the tiniest push.

My therapist wanted to talk about my childhood, even though I told her it wasn't relevant to the situation. In fact, the more I told her it had no bearing, the more she wanted to talk about it.

So I told her about my childhood in clinical detail: the bullying, the obesity, the undiagnosed learning disabilities. After I gave her these details, I presented my emotionless demeanor as evidence I'd healed all those past traumas. She admitted that my recollection was both impressively detailed and impressively detached.

Then she asked me how I felt about those experiences. I immediately started to have a panic attack. I'm not prone to panic—I can generally turn off negative feelings like someone turning off a tap. But this was different, like trying to hold back the pressure of the Hoover Dam with only a bathroom faucet. I sensed an impending breach of emotional pressure that I hadn't even known was in me.

It scared the shit out of me. I felt out of control, like a pilot who discovers that his control stick isn't working.

For weeks we returned to these childhood scenes. My therapist asked me what I thought would happen if I stopped fighting the pressure and just let it out. I told her I would probably sob or get really angry—or possibly both.

"Why would that be a bad thing?" she asked.

"Those are such unpleasant, wasteful feelings."

"Why are they wasteful?"

"We only get so many moments in life. Why would I spend them on feelings that don't feel good?"

"It's not healthy to bottle up emotions, Mike."

"Why?"

"What do you mean?"

"Why isn't it healthy? What's happening in the brain when I do?"

"I'm not sure what's happening in the brain," she said, "but it's widely accepted in human psychology that it's not healthy to repress sadness and anger indefinitely."

I knew there had to be some data behind that observation.

Our emotions are controlled by chemicals that work on our neurons. If repressing sadness and anger was really unhealthy, a neurological mechanism had to be at work.

So I went on a mission to learn the neuroscience behind psychotherapy. It seemed such a waste of time to talk about the same thing over and over or to pay someone to cry in her office for 50 minutes a week. But it turns out that my therapist was right.

Trauma leaves an impression on your brain. And when you recall traumatic events from your past, or when you experience things that remind you of them, your amygdala recreates the fear and pain you had in those moments.

Leaving my church was traumatic, and that trauma was amplified by my childhood experiences of rejection. That sentence, simple as it is, brings tears to my eyes even now as I type these words. My fierce independence, my willingness to walk away from the crowd, was merely a coping strategy I'd developed in response to being bullied. The way I played it, no one could reject me if I didn't need them in the first place. I'd bet a lot of you know exactly what I'm talking about.

But therapy works because human memories, just like human brains, are impressionable. When you recall a troubling memory, you have a chance to modify the deeply emotional responses you experienced at the time of the event. This is why talking about traumatic events in a safe environment or with people who care about you has a way of diluting the trauma and training the brain to understand that the danger is over, that you can be safe.

When I learned this, I stopped trying to hold back the dam. I let the briny river of grief rush through me.

Oddly enough, it made me feel clean.

If you've ever felt hurt by a church, you'll have to grieve that loss to be healthy. The deeper the wound, the more time you'll need. Sometimes sorting out your emotions and reactions will feel like digging through a restaurant's Dumpster without gloves, but all those feelings have to come up—the neurological associations with trauma must be exposed and healed.

The more painful a story is to tell, the more we need to tell it. Jesus spoke a lot about the importance of forgiveness—a crucial teaching but one that has sometimes been used by religious folk to dismiss people's suffering or deny their own. But the science on processing pain, grief, and trauma is clear: When people attempt to shortcut or disavow the sorrow of emotional wounds instead of expressing it, they might unconsciously harbor hostility or helplessness instead of forgiveness. They may experience more psychological harm. Experiments show that our intellectual, emotional, and even physical performance can be affected by emotions denied breathing room.

I advocate forgiving others as a means of self-care, an essential component in both grief and recovery. But that doesn't mean forgetting the harm that's been done, and it absolutely doesn't mean you should ever continue harmful or abusive relationships. What it means is that if you don't process your own emotions and come to forgiveness in a necessarily slow, natural progression, your brain will spend its energy ruminating over that hurt and robbing you of relief.

So talk to trusted friends, see a therapist, hit a punching bag—do whatever you have to do face and process your grief so you can move on without excess emotional baggage.

Despite all my complaints and reservations, I remain hopelessly in favor of being part of a church. My reasons go beyond sentiment. Sociologists and psychologists underscore the importance of community in forming and maintaining a healthy outlook and values.

Humans are among the most socially oriented animals on our planet. We're not especially fast or strong compared to other mammals our size. We have outsize brains, but even that doesn't help much if you're a lone homo sapiens facing down an apex predator. Prior to the advent of weaponry, a solitary human was no match for any of the serious predators in the animal kingdom.

But a collection of humans wielding spears can be nearly as powerful as any creature on Earth. The fossil record shows us this. Our invention of thrown weapons is mirrored by a sudden collapse of apex-predator populations in every region of Earth populated by humans at the time.

Because of our individual weakness and collective strength, evolution has shaped our brains to devote incredible resources to socialization. It's not just a matter of coordinating the hunt. To survive, the earliest bands of humans took on separate roles: Some hunted; others kept watch, gathered food, or tended children. We thrive in cooperation and fail when we're on our own. That's why we're naturally terrified of isolation—for our ancestors, loneliness could mean impending death.

Thus, humans have an incredible incentive to behave in ways that secure a place in their tribe, and spirituality itself played a significant role in the social cohesion and governance of the earliest human societies. We humans are primed to believe what those around us believe, and since our beliefs often drive our behaviors, we have a great incentive to hold values consistent with those of our tribe—a process sometimes called groupthink.

This phenomenon is in no way relegated merely to the distant past; modern humans are equally obsessed with social conformity and status. Witness the way we idolize Kim Kardashian or Brad Pitt—our brains see in these famous individuals all the signs of high social standing, and we want to be associated with them. You might hate "networking," but it's one of our most ancient traditions.

Witness, too, our human tendency to clarify our identity with group labels. Liberals hold in common certain beliefs and values regarding human society and government, and the same is true if you call yourself a Republican, an atheist, a Christian, or even a golfer.

This tendency is so powerful, it can actually cause us to unconsciously filter out and dismiss any information that contradicts our chosen group's identity or norms. One of the reasons so much tension exists between differing social groups is that we humans have a hard time processing information that could challenge or undermine our group's identity.

This matters a lot if you're someone who wants to believe in God but harbors serious doubts. If you truly want to believe, one of the most powerful things you can do is spend time with other people who do, letting them help imprint your brain with their experience of spirituality. People who doubt often leave the church—sometimes against their will—and, end up spending less time with people who believe in God. If you want to know God, it turns out some advice my grandmother gave me mirrors what science has to say: Pray, read the Bible, and go to church!

This capacity of the group to help people form and maintain a belief in God and to experience the positive health and emotional effects thereof forms the basis of my claim about what a church can be and do and why that is good for the world:

> *The Church is at least a global community of people who choose to follow the teachings of Jesus Christ. Even if this is all the Church is, the Church is still the largest body of spiritual scholarship, community, and faith practice in the world—and this practice can improve people's lives in real, measurable ways.*

Belief in God can be beneficial to people, and prayer can rewire the human brain. The largest institution in history devoted to belief in God and exercising that belief is the Church. The scale and scope of the Church is a vast resource for anyone who wants to know or experience God. Beyond its more numinous activities, a church is a ready-made system of social support, community, and belonging.

But what if you don't have a church anymore? How do you find a church that works for you, especially if a church has hurt you in the past?

Taking some time away from church proved healthy for my family. We all had grief to deal with, and our lazy Sundays provided an essential time for decompression. We had to cope with what was before we could begin looking for a new church home.

On the other hand, I can see why some people stop going to church and never return. A church doesn't have to be a building with pews, a choir, a pastor, and Sunday-night socials. That's the church many of us have known, but it's not the only way to seek contact with God among like-minded people who will know, love and accept you. So why worry about getting up early on a weekend morning, dressing up, and driving to a steepled building, when community can be found in other ways?

I don't have a good answer for that. I really don't. I just realized at some point that *my* spiritual life was linked to the weekly rhythm that a Sunday-morning worship service provides. I longed for a time and place set aside for worship, to sing hymns with other people, and to hear a few wise words from a pastor. I wanted the same for my family.

Here's the thing: In my opinion, secular life doesn't provide the kind of ready-made community that a church does. Studies tell us that people are increasing isolated and lonely in the Western world. We're increasingly wealthy and mobile but far less connected to our neighborhoods, civic groups, and other social organizations. We have thousands of Facebook friends but no one to spend time with. In the modern world, a church community offers one of religion's most powerful and least understood advantages.

And pastors—wow. A pastor is someone who will visit you in the hospital, celebrate your new baby with you, and sit with you when you lose a loved one. A pastor's door is open when your marriage is on the rocks or when you're torn over a major life decision you have to make. A pastor can be a wise counselor, mentor, and friend, and when my family left our church, I didn't have one anymore.

That's why I couldn't be a permanent member of the Sunday-brunch society. Of course, even after taking the time to grieve and heal, I found the thought of finding a new church daunting. I'm a total theological weirdo now, but I was even further our in left field three years ago. I never wanted to be a source of controversy again, and I dreamed of a church where I could be next to invisible.

We tried a few churches, but none really fit. There were churches that seemed open but were either very small or very old, and it was important to us that our kids could spend time with other kids at church. We didn't want them simply to come along for the ride—they had to be a part of the community themselves.

Every church we found that was big and full of young families like ours also seemed to hold to the kinds of strict biblical values that had pushed me toward the exit at our old church. I almost gave up.

But a friend of mine told me about a church on the south end of town that she'd visited and found to match the criteria I was looking for: It was multiracial, multicultural, and affirming of people of every race, gender, and sexual orientation. I asked how the church related to women in leadership, and my friend told me that the pastor and most of the staff were women.

That church, Good Samaritan United Methodist Church, became our new church home. It's a safe place for us, where we've been able to grow and to serve. But I'm not telling you about my church because I love it (even though I do); I'm telling you because if there's a church that accepts me, a nontheistic, mystic follower of Christ, then there is a church that will accept you, too.

And that's key. Not every church is for every person. I've got friends who would be really uncomfortable at Good Sam. Despite its unself-conscious orthodoxy, Good Sam is the kind of place where theological and moral questions are welcomed, and people can participate without knowing what they believe about anything. It's theologically Christian, but it's almost Unitarian in its radical inclusion—and that doesn't work for everyone.

I've come around to the need for multiple church denominations for this reason: The fact that Catholics, Baptists, charismatics, and Methodists don't have to share the same space every Sunday may well keep them from strangling one another. The myriad of Christian denominations is more blessing than curse in other ways; it means there's almost certainly an entire stream of Christianity you've never tried that could be perfect for you.

When it comes to finding a congregation you can serve as a part of, there are two things you have to look for: a church that is safe and a church that will challenge you. You should find a church that can share or accept your views on evolution, same-sex marriage, social justice, and environmental concerns; that's part of what makes it safe. Your church should affirm you and accept you exactly as you are, should celebrate how you were made and how you've grown, and should tend to your wounds and love you as you heal. But it can't stop there.

Your church also has to challenge you to become all you can become. It should comfort you, but it shouldn't let you get too comfortable. The people of your church should challenge rote thinking and decision making and prompt you to put your ideas into loving action—to embody the Gospel with hands made dirty by work in the world. The congregation should empower you to serve the world with grace and to see that world with ever-more-loving eyes.

I'd go so far as to say it should make you become more like Jesus, but don't tell anyone I said that.

CHAPTER 14

The Good Book

My great-grandmother gave me a Bible the day I was born. It's a serious tome: Set two bricks next to each other and then top them with a third, and you'll have something about the size and weight of my Bible. Its garnet leather cover is worn thin in places, and the metallic letters on its spine are fading. My name, however, embossed in gold on the cover, looks nearly as fresh as the day I got it.

Like that of any proper Baptist family, this is a King James Bible, complete with study notes. It's a Bible that means business.

When you open the front cover, the first thing you see is a "gift" page. My Bible was a gift from a maternal great-grandmother, but it's my dad's hand that wrote my name, the date, and the occasion: *Birth.* With obvious difficulty, my great-grandmother signed her name, along with a Scripture reference: *Rom. 8–28.* (For those of you who didn't grow up doing Bible drills, that verse reads, "And we know all things work together for good to them that love God, to them who are called according to his purpose.")

My parents kept the Bible in a box at the top of my closet until I was old enough to handle it responsibly. I think my Mom was scared I'd ruin it with markers or leave it outside in the rain. But for some reason I craved its gravity, so I'd beg my Mom to let me take it to church instead of my more durable kid's Bible, a request that wasn't approved until I was in elementary school. It was only then that I realized just how heavy it was.

It's remarkable how well this physical book represents my family's approach to the Scriptures. I had, from the moment of my birth, a serious Bible, touched by both sides of my family and held in high regard—protected, honored, valued. I spent hours and hours poring over its pages, though not with enough regularity to avoid a nagging guilt that I didn't do so often enough.

At first, the colorful maps in the back were the most fascinating pages to me, especially the intricate diagrams illustrating the changes in Jerusalem's layout over the years. Around the time I learned that girls didn't have cooties, the Song of Solomon became my (secret) favorite book. By the time I learned to drive, I'd become fascinated with the introductions to each book that explained who the author was and the historical context in which the book was written. I even read a section called Between the Testaments, which was sandwiched between the Old and New, describing the historical events that had happened between these two halves of the Bible. Quoting this section was a guaranteed way to impress and possibly overwhelm your Sunday-school teacher.

I've owned a lot of Bibles over the years, but this was always my "real" Bible. When I was wrestling with serious theological or ethical quandaries, my New International Version would find itself playing second fiddle as I heaved my King James Bible from its shelf. Its thin pages were even my first resource for my scientific explorations, as well, as I scanned Genesis to learn more of what it had to say about our human origins. From Paul's writings, I learned how to be a good husband and a good employee. And the Bible was the place where I could follow the footprints Jesus left on Earth, a place where the Son of God took on discernible form. I trusted the Bible to teach me how to love and serve God and how to navigate the challenges of life.

I loved the Bible with an enduring sincerity I've never held for any other book. That's probably why I felt so betrayed by it in the years following that infamous family meeting. More than anything else, it was reading my Bible that had turned me into an atheist.

When I came back to faith, I made peace with Christianity one piece at a time. First, I found scientific terms for the forces and experiences humans call "God." Those insights made me comfortable leaving the word *atheist* behind. Prayer came next, a practice science revealed to be tremendously valuable, and once I could pray unself-consciously, it was a lot easier for me to feel like a real person of faith instead of a facsimile. Coming to terms with Jesus was more difficult, but admitting the limits of my knowledge helped me, and finding a healthy church for me sealed the deal.

Somehow, though, I couldn't quite get back to the Bible. I tried, of course, even going so far as to pull that leather King James Version from the shelf to look at the maps, trying to recapture the wonder I'd experienced as a child. But it didn't matter what Bible I tried, or even what book of it I turned to. I couldn't read much—at most, a page—before I stumbled into a phrase that knocked the Bible from my grip.

For example, one of its books I had enjoyed most while growing up was Proverbs. It's a collection of wise saying and advice, and I often read it for guidance on how to live. But looking at Proverbs from the other side of atheism revealed the book in a new light. The "Wisdom of Solomon" tells parents to beat their children, even to ignore their cries as they do so.

Then there's 1 Corinthians. It's a book ascribed to the apostle Paul, who says any woman who refuses to cover her head in church should have her head shaved. This is just after he says that any man who has long hair is shameful, while long hair is a "glory" to women. In general, Paul's anti-woman stance troubled me and stood in contrast to a Jesus who spoke first to women after his Resurrection.

In too many places, the Bible reflected an anachronistic, absurd view of science and/or history. Genesis still spoke of a single male and female being the origin points of humanity, and the Torah still equated God's favor with military conquest. Job still spoke of a God who makes wagers over a man's life and who allows his family to die in order to prove a point. Despite compelling passages about love, forgiveness, justice, and peace, I couldn't shake the sense that this was a book full of primitive brutality.

This might sound strange to some people, but I was actually afraid to read the Bible. Every time I did, I felt I was losing my faith all over again. Nosing through Genesis or Jeremiah imparted God with a fairy-tale quality that I sometimes couldn't shake for days.

So I finally stopped trying. I had read the Bible so many times that I was able to keep up with most conversations about it without reading up first; I could participate in Bible-study groups without actually rereading it. My faith seemed easier to maintain if I stayed away from the Scriptures—those words were too associated with my loss of faith.

But before I lost my faith, the Bible was a book unlike any other to me. When I read the Bible, I felt a part of something: a lineage of people who'd followed God for thousands of years before I was born. The Scripture placed me inside a movement that transcends cultures and historical eras. I had to find a way to read the Bible without either crossing my fingers or waking my most fiery skepticism. But how?

What, if any, value does the Bible have for people today? It's a question I couldn't answer for a long time, but I've been on a mission to answer this question for myself. When I found the answer, it transformed the Bible from a curious historical artifact into a page-turner that I can't put down.

Over the last two years, I've learned that my problems aren't with the Bible at all. All the anachronisms, contradictions, and similar stumbling blocks I found in its pages aren't flaws in the Scripture. Instead, they are flaws in the assumptions I hold as I read the Bible.

I had long assumed that the Bible was a single book written by a single divine hand through many men, meant to contain God's perfect and complete revelation for humanity. That's not entirely my fault—I was handed that lens from my earliest days on Earth.

I'm not alone here. This is the way many churches teach the Bible, and it puts responsibilities on the text that are impossible to fulfill. A book authored in this "divine" way would have to be completely free from factual error or contradiction. It would have to be *perfect*. It must clearly communicate divine will not just for the time in which it was written but for all time to follow.

This is why many atheists joke that the most powerful tool they have in making new atheists is the Bible itself. No book can meet such impossible expectations, and many believers have had their faith wrecked on a reef of biblical criticism.

Many Christians have felt that they must either to accept the lens I was given at birth or dismiss the Bible entirely. This couldn't be further from the truth. In the last two years, I've found that both historical and contemporary biblical scholarship is broader than I ever imagined it to be. It turns out that people have been wrestling with how to handle the Bible for a long time—including the people who assembled the Bible in the first place!

The Bible seems contradictory because the Bible *is* contradictory. After all, the Bible isn't a single book, with one voice, one perspective, or one unified take on the history of how God has interacted with people. This should be obvious, after all. The Bible is divided into *books* first, not *chapters*. It's more like a library than a book—and, for that matter, the Bible isn't even one single library. The Catholic Bible has 73 books, while the Protestant has only 66. Protestants assemble the Old Testament differently, mainly because Martin Luther stripped six books out in the 16th century.

I mention this not to be difficult, but to underscore an important point about the Bible. Countless authors, scribes, editors, councils, and translators have all imparted their perspectives on its text. Every book in it was selected for inclusion by people, and people haven't always agreed about what books should make the cut. There was no single council or person who selected the list, and the process was political, heated, and controversial.

But where did these books come from? That leads us to a second important fact. These books were written by different people over an impressive span of time—1,500 years. Each book was written by a specific author, for a specific audience, with a specific agenda. Historians believe some books were surreptitiously edited over time, often with political motivations. This is especially true for the Torah, the first five books of the Bible. Further, modern scholars believe that many of the books' authors aren't who the Church has traditionally said they were.

In his book *The Bible Tells Me So: Why Defending Scripture Has Made Us Unable to Read It,* Bible scholar Peter Enns characterizes the Bible not as a single, inerrant text, but a chronicle of an ongoing conversation—even debate—about God. Far from being accidents, the contradictions found in the pages of Scripture are often intentional, reflecting the different motivations and opinions of the Bible's many authors.

This isn't just an Old Testament thing; the New Testament does this, too.

The author of Matthew writes to a primarily Jewish audience, with an emphasis on telling the story of Jesus with parallels to the story of Moses. Moses, of course, is the major figure in Jewish Scriptures, and painting Jesus with a Moses brush was good storytelling.

The author of Matthew (who may not have been Matthew or even a disciple at all) starts his Gospel with a *lineage.* His first move is to establish Jesus in the context of Jewish history, starting with Abraham, and with a none-too-subtle inclusion of King David.

If you didn't grow up in Sunday school, this information may be lost on you. Matthew starts out by putting Jesus on an all-star roster of superstars from Jewish history. This Gospel also takes painstaking effort to point out every detail of Jesus' life that lines up with Hebrew prophesies about the Messiah. It says nothing about mangers, shepherds, or angels singing to announce the birth of Christ.

Even so, it says loads more than the book of Mark. Mark skips from a pre-birth prophesy to Jesus' adulthood without ever mentioning his birth or childhood. Of all the Gospels, Mark's motivations are the hardest to read, and its telling of the life of Christ tends to treat Jesus' divinity as a carefully guarded secret. Mark may be the second Gospel in the Bible, but most scholars think it's the oldest of the four, and many also think the earliest versions of Mark didn't include the Resurrection.

Then there's the book of Luke. The author of Luke likely wrote Acts, as well, and seems to be well educated. He's the first Gospel author who expands the scope of his audience to include Roman citizens and non-Jewish audiences. Luke uses literary devices and images that would be well understood by a Roman audience—including callouts to Christ's divinity with singing angels and a special star in the sky.

The "black sheep" of the Gospel authors is whoever it was who wrote John. It represents a radical departure from the other three books, with a Jesus who speaks in long monologues and is crystal clear about his divinity when contrasted with Mark's guarded Christ. Where Luke expanded the audience beyond Jews to include Romans, the author of John consistently paints Jews as the enemies of Jesus—even though Jesus himself was inarguably Jewish.

When I talk about the Bible with primarily Christian audiences, people nod when I say the Bible is not a book, but a library of books. Heads keep nodding when I say it's also true that the Bible was written by many people. But when I bring up the third point—that the Bible contains multiple, contradictory perspectives—it tends to inspire an odd energy in the room, something between anger and panic.

I'm not the only person who carried a lot of assumptions when I read the Bible, and it can be tough to entertain the idea that the Word of God has different perspectives in it. Biblical apologists spend all their time weaving these different viewpoints into a single frame, in an effort that often looks like squids playing Twister: fascinating, appalling, and hard to follow. We've seen what this approach to history can sow: a destructive oversimplification of the Church's past.

Americans treat their national narrative in much this way, too. We simplistically teach a single story in our history classrooms, of brave rebels who left cultures of tyranny and heroically crossed the Atlantic to found a nation built on freedom and justice. When we speak of our national sins, such as the genocide committed on Native Americans or the brutal, long-term economic extraction of wealth from black bodies via slavery and segregation, we seem to dismiss these troubling matters as things that happened in the remote past but that have been solved today.

We often tell ourselves the easy story, not the messy, multiparty conversation required to view our national history in its true light: complex, contradictory, sometimes cruel, and never quite resolved. Is it any wonder, then, that we tend to read the Bible in this way?

Fortunately, respected Christian scholars have taken the time to address this. British New Testament scholar and former Anglican bishop N. T. Wright, who's kind of the Stephen Hawking of theology, often speaks of the "Word of God through words of men." Peter Enns draws parallels between Jesus and the Bible, calling both "fully human and fully divine." Speaking on the nature of the Bible, Rob Bell says, ". . . whatever divine you find in it, you find that divine through the human, not around it."

These are comforting ideas I see value in. They're also nearly impossible to verify empirically and, therefore, subject to the same acidic erosion via doubt that most theological ideas are.

But here's the thing: Even when I approach the Bible via secular scholarship on archaeology and anthropology, not theology, I still find a book worth reading. I don't believe in a literal Adam, Eve, or serpent, but I do know what's it's like to hear the whisper of that serpent while looking at a forbidden fruit. I doubt that Earth was ever submerged under a global flood, but I do understand what it's like to feel that evil is all around, running rampant in the world. Like doubting Thomas, I know what it's like to encounter a savior but demand to see the wounds before I will accept the miracle.

When I let go of the Bible as an inerrant document and embraced it as a multi-party discussion about God, all of a sudden, I began to see a book I could appreciate on its own terms. That's important. After all, this is the guiding document for the world's largest religion—a faith I happen to be part of.

I've learned to approach the Bible again by making use of a secular understanding of its text, while allowing the religious response to its contents to direct my faith journey. Stated as an axiom, my stance reads like this:

The Bible is at least a collection of books and writings assembled by the Church that chronicle a people's experiences with, and understanding of, God over more than a thousand years. Even if that is a comprehensive definition of the Bible, study of Scripture is warranted to understand our culture and the way in which people come to know God.

For me, it's liberating to look at the Bible this way. It's OK to acknowledge that the Bible's contradictions of science reflect the pre-scientific understanding of the universe held by its authors. The brutality recorded in Scripture, such as Israel's conquest of the promised land, along with its oft-quoted command to kill not only the men, but women and children, as well, reflects a time when one's ability to win wars was considered a reflection of divine favor—and anthropologists' accounts of the actual historic events reveal a reality far less dramatic and violent than what lies in those pages.

In the Bible of human authorship, I see Job wrestling to understand the tragedies of his life in the context of his love for God. I see Paul wrestling with his murderous past in the wake of that blinding light on the road to Damascus. I see Moses with one foot in power and the other in slavery, with a longing to simply run away instead of speaking a dangerous truth and taking the responsibility for others that come with it. Every one of these stories is more powerful, not less, when viewed as a document penned by humans.

The Bible should be read in the same way we'd read any work of ancient literature. No one mocks Plato for his celestial spheres, even though modern science reveals them to be false. Instead, we put that idea in the context of history and see how it contributed to humanity's growing understanding of the world.

Ancient literature must be read in a historical context, and the Bible is no different. But there's more to a more open reading of the Bible. Dropping our assumptions also lets us glimpse something far more fundamental—that the Bible isn't science or history. The Bible isn't a legal document.

The Bible is art.

Ask someone to tell you about Vincent van Gogh, and most people will tell you he was a painter who cut off his ear. That's tragic, because it distills his fascinating life down to a single, sensational fact. It's as if the most remembered piece of Steve Jobs's legacy was that he had a troubled relationship with the mother of his first daughter. It's a fact, but it doesn't capture the story of the man's life.

What you may not know is that van Gogh wanted to preach the Gospel and actually sought to become a Methodist minister. He was a man of deep faith who wanted to serve others, and his experiences as an art dealer left him cold to commercial enterprise. His zeal for God, however, didn't translate into academic achievement. He failed both an entrance exam to a school of theology, as well as a three-month course at a Protestant missionary school.

But van Gogh didn't give up. Ineligible to become a pastor, he took a missionary post in a small Belgian village. I guess van Gogh had read the New Testament, because he chose to eschew wealth for a very simple life. Soon after arriving, van Gogh sold his possessions and began sleeping in a haystack behind a baker's house.

Imagine van Gogh showing up to preach, covered in hay and smelling like bread. His congregation didn't find this visual anywhere near as fun as I do—they took a collection to raise funds and told Vincent to buy a place to stay and some clothes that didn't smell.

But Vincent van Gogh did a remarkably Christ-like thing: He gave all his money away to the poor coal-miners in the village. He returned to his haystack, where the baker's wife reportedly would hear him sobbing at night, possibly grieved by the poverty all around him. But as much as Jesus might have approved of his actions, neither van Gogh's congregation nor church authorities could cope with the idea of being led by a homeless man.

So they fired him.

The reaction of Vincent's family was mixed. His father often spoke of committing him to an insane asylum, but his brother Theo told him to pursue art. Theo went so far as to support Vincent financially. Vincent taught himself to draw and to paint, and he became an artist. Having been disenchanted by the commercialization of art and disenfranchised by the rejection of the Church, this painter who loved Christ learned to reveal the tragic beauty of human existence on canvas.

When you understand this story, you can look at the profound beauty of van Gogh's work with a fuller appreciation. I've stood in front of *The Starry Night*—the actual work. It's as much a sculpture as a painting, with gobs of paint piled thick on the canvas in a way that creates rich textures and depth. No photograph does it justice.

Somehow, van Gogh's masterpiece reflects a kind of light that is more than real—it shows light and dark sharing space, blending and influencing each other. It affects you in a way that's both beautiful and haunting, joyous and sorrowful.

At the center of the frame is a village; lights are on in most of the windows there. A church is prominently featured, but I can't help noticing that its windows are dark—something my eye never caught before I heard the religious part of van Gogh's story.

When I look at *The Starry Night,* I am, through some miracle, given a gift: the gift of seeing the world through another person's eyes. Without a single word, written or spoken, I am presented with what might represent the soul of Vincent van Gogh. I can see his hope and his despair, his joy and his grief. I see a world of aching beauty and darkness.

Van Gogh's view of the world becomes a lamp that reveals corners of my heart that I didn't know were there—and all of this happens immediately, even though he died 88 years before I was born.

So, ask might yourself this:

Is The Starry Night infallible?

The question doesn't make sense. Though grammatically sound, it is a query with no meaning. I could just as easily ask, "How much does a sunset weigh?" The beauty of *The Starry Night* isn't in its being fallible or infallible. It's a window into another person's soul.

Let's try another question:

Is The Starry Night true?

If we're talking logic or math, this question is as nonsensical as the first. But if we ask with the perspective of an artist or philosopher, we might find that, yes, *The Starry Night* is very true—it tells us truths about the human experience. It's a testament to how grief feels and the numinous quality we often experience when we peer deeply into the night sky.

The Starry Night shows a vast sky, overpowering from our

point of view, and at the same time it tells you what it's like to have a broken heart. It is somehow more true than facts—it resonates in some deeper chamber of the human heart.

So let me ask you two more questions:

Is the Bible infallible? Is it true?

We're asking the wrong questions about the Bible. Accepting the Bible for what it is, a library of books written by men about God, seems at first like a profound demotion for the Good Book. But I believe this approach solves the problems doubters face when they approach the pages of Scripture—all without evicting God from those pages.

I wrote a song for my wife once. In an unprecedented stroke of creative mastery, I titled it "Song for Jenny." This wasn't just any love song. I was getting ready to ask Jenny Frye to marry me, and I wanted her to feel as if she was in a romantic movie.

I bought as many candles as I could possibly find to light all around our church. I got rose petals from flower shops and scattered them down the central aisle. I called Jenny and asked her to come pick me up from band practice.

When she walked in, the band started playing, and I sang her the song. When I stepped offstage, the song continued, and I'd even written another chorus for after she said yes. I wrote "Song for Jenny" because I loved her more than anyone I'd ever known. As the title not so subtly suggests, "Song for Jenny" was inspired by Jenny Frye.

If someone asks me if God inspired the Bible, I say yes without hesitation. I think the Bible was inspired by God in the same way "Song for Jenny" was inspired by Jenny. She didn't write it—what a strange notion—but you can learn a lot about her if you hear it. You learn just as much about how I feel about her and what our relationship is like. You may even catch higher themes about what it's like to be in love and the value of marriage.

You can't hear "Song for Jenny" without pondering its subject. And you can't read the Bible without encountering God.

The value of the Bible is revealed in that way. It's the stories of people like me, and maybe yourself, who have looked up at the night sky and searched for God. Their achievements and their failures are there. There are stories of transcendence and war, grace and exploitation—stories that reflect an unfolding drama as broad as the human condition. I find myself in its pages; I find solidarity in the story of doubting Thomas and the other disciples as they stumble along behind Christ.

I'm at least as self-contradictory as the Scriptures. Some days I feel that Jesus walks beside me, and other days I feel I'm completely bonkers for buying into the Resurrection. I've denied Christ more than three times. But when I read the Bible, it gives me confidence that even I, with all my hang-ups, am included in the sprawling, multi-millennia story of God's people.

I no longer fear the Scriptures. I'm back to reading them every day.

CHAPTER 15

Finding God in the Waves

"Why do you even bother, Mike?"

I was on the phone with an old friend from California who said I couldn't use his name in this book because he doesn't "want to be stalked by a bunch of God-nuts if your book takes off." He's a very serious atheist, the kind who wears Flying Spaghetti Monster T-shirts and tweets jokes about how crazy it is to believe in God.

We'd been talking about some frustrations I was experiencing. Some prominent Christian websites had recently called out one of my podcasts as being dangerous, even "more dangerous than atheism." A few months prior, some old friends had taken me out to lunch to rebuke me "in the name of Jesus Christ," while demanding I shut down my own website.

I was feeling pretty down, and my friend called me out of the blue to give me a chance to vent. I took it.

His question startled me. I didn't have a good answer, so I played dumb to give myself time to think. "What do you mean?"

"Why do you bother trying to put intellectual legs on this hokey, Bronze Age mythology?" he repeated. "I've read your stuff, and it's not entirely bat-shit, but you can't think the God of your axioms has anything to do with the God of Christianity, that nut Yahweh who demands blood sacrifices to appease his emo angst.

"What do you think you're going to get from churchgoers? A hero's welcome? An unpleasant lunch and some angry blogs are just the beginning, my friend. All they're going to do is spit in your face.

"Admit it, what you believe is a lot more like humanism than Christianity. You're using ancient language to describe modern ideas and then pretending to be a part of the ancient tradition. But you're not. I know it, and the crazy-ass Fundies know it, too. I'm just waiting for you to realize it—you'd make one hell of an atheist blogger. I'm living proof that atheists come across as angry, but you wouldn't. You've got all that 'nice guy, listens to everyone' bullshit. People would eat it up."

That's the clean version of his speech to me on the phone that day. And although my friend meant well, I have to be honest: His words rattled around in my head for a few days. He had put his finger on something I'd always feared.

When I first came back to faith, I didn't call myself a Christian. The word was just too loaded and carried too many assumptions. *Christian* can invoke some troubling historical baggage: colonialism, imperialism, systemic and personal racism, sexism, homophobia.

Then there's the not-so-small matter that it requires a very broad take on Christian orthodoxy to place me inside Christianity. More open-minded Progressive Christians are happy to do so, but I can't sidestep the fact that a significant number of American Christians wouldn't say that I'm a Christian at all. I try to avoid self-labeling as much as I can, for the sake of reducing my own neurocognitive biases, but sometimes you just want to be able to answer the question "What religion are you?" without vaulting into a speech.

I couldn't comfortably wear the label *Christian* again until I found a church—an honest-to-goodness mainline American Protestant church—that would count me among its members. When I did, it gave me an easy answer to the accusation, "How can you call yourself a Christian at all?" I now say, "I'm a member in good standing of the United Methodist Church."

But that doesn't get to the core of what my friend was asking: Why do I care about being a part of the church or using such words as *God*? Wouldn't life be simpler if I just said "universe" instead of "God," or "transcendence" instead of "spirituality"? I admit, in many ways, it would.

But what about that indescribable light I saw on the beach, or that voice I heard at the Communion table? What about the stirring in my heart when I was seven, or the way people somehow can give voice to difficult truths when they sit together with heads bowed—truths that would otherwise be too hard to utter?

What about that echo, still ringing from the shore of the Pacific, a feeling that God is all around me? A feeling that when I lean into it, I see the profound beauty and value of every person and am given eyes to see the suffering of people that wouldn't otherwise affect me. Eyes that see the scars of the lesbian whose parents won't let her bring her partner home for the holidays, or the black teen who's a suspect or, worse, a casualty, for walking the same streets I do?

That's the point. The way I see God is not an attempt to take the mystical out of reality—it's a way to incorporate the miracle and mystery of the life we can see, measure, observe, and test every day. A world composed of countless strange little particles, themselves made of energy and invisible fields that defy the imagination. Skeptics challenge the idea of an unseen spirit realm, but what we know about the physical realm is far more fascinating and strange than a bush that burns without being consumed. We're already numinous and ethereal, beings made of mostly empty space and probabilistic waveforms.

So, yes, I sometimes use new metaphors for *God,* blending the words of the ancients with the insights of our times. But it just so happens that doing so plants me deeply within the biblical Christian tradition.

Because first God was with us in a garden, and we could walk and talk with him just as we could with anyone else. This was an easy way for us to understand God, but in time we grew to understand that this was too limiting a way to talk about our Creator.

So God became a bush that burned but was not consumed, who described himself with such sweeping language as "I am that I am" and "I will be that which I will be." Then God became a pillar of cloud by day and fire by night, to show that he was the God of the people of Israel, guiding them through their exodus and exile.

Then God was in a ceremonial Ark of the Covenant, a physical symbol of his Word, carried onto the field of battle to offer his favor to his people, and then he dwelled in a temple, anchored like other gods of that age to the soil of his people.

First a man, then a bush, then a pillar, and finally a spirit that dwelled in a temple in the greatest city of his people's nation. God, a great mystery, accessible only through priests and elaborate rituals—a powerful figure, but one not easily understood.

So God once again became a man, a human with a face, who could walk and talk with us like anyone else. God told us stories and parables, and said "blessed are those," offering a picture of human life that overturned the one we had made ourselves.

This was too much for us to handle, so God hung on a cross. A weak, broken God cried out, begging that we be forgiven before saying, "It is finished."

God became an empty tomb and a man who could walk through walls. A holy, infinite Father and a Risen Son. The Son left Earth, and a Holy Spirit was sent instead, as mysterious as that being in the temple, only now the temple was each of us.

From a bush to a pillar, to an ark, to a temple tied to soil, to a man free to roam, to a Holy Spirit who dwells in the hearts of all God's followers. That's God at the close of the Bible story, but God continued to change.

God became a Trinity to the early church: Father, Son, and Holy Ghost.

Which one of those is the real God? The genuine article? They *all* are, of course. All were ways to describe God appropriate to a time and place, and all were a reaction to some earlier image. Times of exile or alienation—dare I even say any time of doubt?—often provide fertile soil for a new image of God to spring up. When that happens, our faces light up like that of Moses coming down from the mountain, and we're ready to tell the world we've done it, we've found the *real* God, the *true* God, the *only* God.

But we've got to stop this. It's not helpful. The God in my axioms isn't superior to the God I once found in the Southern Baptist faith and message. Somehow, over time, we humans seem to find the image of God we need in order to serve and grow and face that often challenging task of existing as a conscious entity. I'm done saying I've found the right one—mysticism tells me that these are all metaphors, all symbols, pointing to a single God who is beyond anything I will ever be able to imagine.

Be it Moses' burning bush or Carl Sagan's cosmos, both propel me to a posture of worship: an understanding that I did nothing to get here, on this planet at this time and with these people, and yet I get to enjoy it all. Every sunrise, every breakfast at the table with my kids, every skinned knee, and every kiss from my wife. Every song, poem, and, yes, every loved one I lose is a gift. To share the joys and sorrows of my friends, to see little ones born and old ones die all tie me to an incredible cycle of unspeakable beauty that I am a part of, and the only possible word I have for it all is this one:

God.

I keep finding God in the waves—the waves of the Pacific, the waves of gravity, the waves of electromagnetic energy, and the waves that move through our brains. I find God in the sound waves of ancient hymns, of children laughing, and in the quiet sobbing of those who say under impossible assault, "I can't breathe."

This is, of course, all wildly unscientific, wildly imprecise. It has to be. I can describe in precise scientific terms what happens when my young daughter grabs my hand without a word before crossing a street. I can tell you about electron boundaries and how, although no real contact occurs, photons and other particles act as force carriers between incalculable atoms.

A click of the microscope up, and I can tell you how I'm a collection of cells, and my daughter is a collection of cells, and when our hands touch, electrochemical signals travel along nerves up my arm, over to my spine, and then right up to my brain, where an electrical storm ricochets through billions of neurons, which in turn produce an incredible chemical soup.

I can tell you these beautiful, miraculous things, and they will tell you absolutely nothing about how it feels to be a dad who loves his daughter and is grateful for her trust. Though science can deepen an appreciation for this fact, it can't convey the overwhelming beauty of that moment.

Only a poet or a painter can do the work of sharing this truest of all things. Love.

I'm finished trying to let my faith, my theology, my reading of the Bible trump humankind's crowning system for uncovering facts about the physical world. I'll never do it again. There is absolutely nothing as effective for learning about physical reality as the sciences, and I love them for it.

But my faith gives me something else. A sense of meaning, belonging, and purpose in the midst of all those facts. It gives me faith that all things will work out for good, that love is the basic reality of our existence, because God is love. These ideas don't have tremendous empirical merit, but they change my life when I hold them in an open mind.

Science gives us fact. Faith gives us meaning.

These two lenses, so often set up in opposition to each other, are most powerful when used together. Somehow, life becomes more clear—and dear—when I refuse to water down one stance for the sake of the other and, instead, dive deeply into both streams of experience and feeling, collecting the truth that flows from each.

We don't have to choose one or the other. Beauty and mystery surround us in every moment. They're easy to miss and easy to crush in the grip of our desire to control them.

But if we open ourselves up to receive both, we'll be surprised by what we find.

And God will meet us there.

AUTHOR'S NOTE

Dear Reader,

Though this book is an honest telling of my life's experiences and how they affected me, it's not a strictly factual or perfectly chronological account. I'm sure my recollection of many events is flawed, and I've also intentionally modified some of my story for narrative flow, book length, and to protect the privacy of some people who are part of my life. I've done my best to be as accurate as possible, but I have sometimes simplified or modified for clarity the sequence of actual events.

That's the easy part. Here's the hard part:

I refer to God as "Him" throughout the manuscript. Much of this book is told in the past tense, and at the time of those passages I thought of God only as "God the Father." Today, I don't think of God as having a gender at all. *Father* was a culturally appropriate metaphor for God in the time the Scriptures were written, but even the Bible contains feminine images of God. Gender is an earthly idea. It's not innate to the universe, much less to God.

As the story progresses, I intentionally used male pronouns for God less. It's a hard habit to break, though. English, among others, is a tough language to use without invoking gender. We don't have gender-neutral singular pronouns. If you write in a way that is more accurate about God's gender, you also make God less personal. It's an unsolved linguistic quandary. If you'd like to see a more accurate depiction of how I relate to God and gender, please check out God Our Mother by The Liturgists.

Peace, love, entropy,

Mike McHargue (known on the Internet as Science Mike)

ACKNOWLEDGMENTS

If it weren't for the love and acceptance of my wife, Jenny, and my mom, Ruth, this would be a book about how secular humanism can succeed where religion has failed. I need to thank them first and most.

I am so grateful for my dad, also named Mike McHargue, for a lifetime of love, support, and encouragement. It's unfortunate that you all met him at his worst in this story, because he's one of the greatest men I've ever known.

I want to thank Rob Bell for responding to my greatest fears with grace. I would not be a Christian today without his kindness. Rob was also instrumental in bringing this book to life—he invested time in me when it seemed least promising. Rob was the first and most ardent supporter of this work.

I worked out where I stood with God during long runs at night with my friend Jeb. He not only ran with me, but listened graciously. I'd be lost without my friendship with him and his wife, Trueby.

Cathleen Falsani devoted countless hours to my vast, incoherent early vision of the Final Answer to Science and Faith. She offered a better plan: Just tell people your story. She coached me through the darkest days of this manuscript.

My friend Bradley Grinnen believed in me when I did not believe in myself. Bradley's patient coaching shaped not only these pages, but also who I am.

My agent, Christopher Ferebee, is the best in the business. Though his excruciatingly high standards gave me sleepless nights, trying to clear that bar transformed this text into something worth reading. I couldn't have done any of this without his insight and wisdom.

David Kopp and Derek Reed at Convergent Books are the perfect creative collaborators. My heart goes out to them in sympathy—Microsoft Word collapsed under the weight of changes required to turn this manuscript into something publishable. My deepest gratitude goes to them for not only understanding this book, but for having a bigger vision for it than I ever could have. Working with them both has been a joy—an absolute joy.

That experience extends to all the editors, production folks, marketers, and creators at Convergent Books and the rest of Random House. I don't think any author has ever had a better experience than I have with her or his publisher.

My thanks go to Jim Chaffee for keeping me operating in some semblance of professionalism. I'd be lost without his guidance.

I will forever be thankful for Dale Fredrickson, Sarah Heath, Gregg Nordin, Chris Hawley, Sarah Johnston, Jeff Campbell, Ty, and the rest of the Laguna Fifty for being my church when I was without a spiritual home.

Thank you to Michael and Lisa Gungor for not only their friendship and collaboration, but also an invitation to help create a space for others to find room at the table.

Thanks to Rachel Held Evans, Donald Miller, Lisa Whelchel, Ryan O'Neal, Bob Goff, David Gungor, and Jacob and Hae-jin Marshall for being a safe harbor as I made a reckless leap from advertising nerd to author/podcaster.

Betsy, David, and my Good Samaritan family provided a church that loves me as I am and challenges me to become who

God is making me. I thank them for being the body of Jesus to me.

Finally, I want to thank all the listeners and supporters of *The Liturgists Podcast* and *Ask Science Mike*. They amaze me, lead me, and give me hope for our species.

AXIOMS ABOUT CHRISTIAN FAITH

FAITH is *at least* a way to contextualize the human need for spirituality and to find meaning in the face of mortality. *Even if* this is all faith is, spiritual practices can be beneficial to human cognition, emotions, and culture.

GOD is *at least* the natural forces that created and sustain the universe as experienced via a psychosocial model in human brains that naturally emerges from innate biases. *Even if* that is a comprehensive definition of God, the pursuit of this personal, subjective experience can provide meaning, peace, and empathy for others.

PRAYER is *at least* a form of meditation that encourages the development of healthy brain tissue, that reduces stress, and that can connect us to God. *Even if* that is a comprehensive definition of prayer, the health and psychological benefits of prayer justify the discipline.

SIN is *at least* volitional action or inaction that violates human consent or produces human suffering. Sin comes from the divergent impulses between our lower and higher brain functions and our evolution-driven tendency to do things that serve ourselves and our tribe. *Even if* this is all sin is, it is destructive and threatens human flourishing.

THE AFTERLIFE is *at least* the persistence of our physical matter in the ongoing life cycle on Earth, the memes we pass on to others, and our unique neurological signature in the brains of those who knew us. *Even if* this is all the afterlife is, the consequences of our actions persist beyond our death, and our ethical considerations must include a time line beyond our death.

SALVATION is *at least* the means by which humanity overcomes sin to produce human flourishing. *Even if* this is all salvation is, spiritual and religious actions and beliefs that promote it are good for humankind.

JESUS is *at least* a man so connected to God that he was called the Son of God, and the largest religious movement in human history is centered around his teachings. *Even if* this is all Jesus is, following his teachings can promote peace, empathy, and genuine morality.

THE HOLY SPIRIT is *at least* the psychological and neurological components of God that allow God to be experienced as a personal force or agent. *Even if* this is all the Holy Spirit is, God is more relatable and neurologically actionable when experienced this way.

THE CHURCH is *at least* the global community of people who choose to follow the teachings of Jesus Christ. *Even if* this is all the church is, the Church is still the largest body of spiritual scholarship, community, and faith practice in the world—and this practice can improve people's lives in real, measurable ways.

THE BIBLE is *at least* a collection of books and writings assembled by the Church that chronicles a people's experiences with, and understanding of, God over more than a thousand years. *Even if* that is a comprehensive definition of the Bible, study of Scripture is warranted to understand our culture and the way in which many, many people come to know God.

NOTES

INTRODUCTION

The stats on creation beliefs, religious trends, church growth, and America's sentiment toward atheists were drawn from a few sources:

"'Nones' on the Rise: One-in-Five Adults Have No Religious Affiliation," Pew Research Center, October 9, 2012.

Barry A. Kosmin and Ariela Keysar, "American Nones: The Profile of the No Religion Population,", Trinity College, 2009.

Michael Lipka, "A closer look at America's rapidly growing religious 'nones,'" Pew Research Center, May 13, 2015.

Frank Newport, God is Alive and Well: The Future of Religion in America (New York: Gallup Press, 2012).

John S. Dickerson, The Great Evangelical Recession: 6 Factors That Will Crash the American Church . . . and How to Prepare (Grand Rapids, MI: Baker Books, 2013).

CHAPTER 2: BINGE READING THE BIBLE

000 ***Yom* can be translated as a year, or even an epoch, depending on the context.** Check out Strong's Hebrew Lexicon, *yom*, #3117.

000 **that our Universe is very old—roughly 13.8 billion years** This belief is so commonly held in the scientific community, it's tough to cite any one source, but Lawrence Krauss does an excellent job explaining the determining methods and observations in his book *A Universe from Nothing: Why There Is Something Rather Than Nothing* (New York: Free Press, 2012).

CHAPTER 3: A FOOLISH BET

000 **He referenced a series of scientific studies that failed to show any positive effect of prayer on people recovering in hospitals.** This is obviously drawn from Richard Dawkins, *The God Delusion* (Boston: Mariner Books, 2008), but here are the studies he cited:

H. Benson et al., "Study of the effects of intercessory prayer (STEP) in cardiac bypass patients," American Heart Journal 151: 4, 2006, 934–42.

Richard Swinburne, [ARTICLE TITLE] Science and Theology News, April 7, 2006

000 **"Your brain has about 86 billion neurons"** This claim comes from the research done by Suzana Herculano-Houzel, an assistant professor at the Universidade Federal do Rio de Janeiro, cited by Jennifer Welsh, "Human Brain Loses Billions of Neurons in New Analysis," livescience.com: http://www.livescience.com/18749-human-brain-cell-number.html.

000 **Your brain can't move, but it consumes up to 20 percent of your nutrients and 25 percent of your oxygen.** Cited from Marcus E. Raichle and Debra A. Gusnard, "Appraising the brain's energy budget," *PNAS* 99, no. 16: http://www.pnas.org/content/99/16/10237.full.

000 **Henry Molaison, Phineas Gage, and Charles Whitman all are members of a long procession of unfortunate people who've shown us how important our brains are.** I don't recall where I first heard the stories of these men, but I used Wikipedia to get a sketch of their lives, and I followed the references and further reading resources to fact-check what I read and to learn more:

https://en.wikipedia.org/wiki/Henry_Molaison

https://en.wikipedia.org/wiki/Phineas_Gage

https://en.wikipedia.org/wiki/Charles_Whitman

000 **More sophisticated brain imaging technology has shown that people's beliefs about God aren't anything like a spot**

and instead arise from a complex network in our brains. The most influential person in how I understand God and the brain is Dr. Andrew B. Newberg. This claim is based on a summary of his research as presented in his book *Principles of Neurotheology* (Philadelphia: University of Pennsylvania, 2010).

000 **In the brains of atheists, *God* is a noun, a noun no more real than *tooth fairy* or *unicorn*. But believers have a rich neurological network that encapsulates God through feelings and experiences that are difficult to articulate with language.** This is drawn from Andrew Newberg, *How God Changes Your Brain: Breakthrough Findings from a Leading Neuroscientist* (New York: Ballantine Books, 2009).

000 **Your brain is a machine that builds a model of the world by throwing other things away.** As described by Michio Kaku in *The Future of the Mind: The Scientific Quest to Understand, Enhance, and Empower the Mind* (New York: Doubleday, 2014).

000 **There's a saying in neuroscience: "neurons that fire together, wire together."** More than one neuroscientist claims this quote as his or her own, which I find hilarious.

CHAPTER 4: SECRET AGENT MAN

000 **Atheists tend to be more intelligent, more educated, and they tend to be devoted parents.** Figures on atheists' IQ were pulled from Miron Zuckerman, Jordan Silberman, and Judith A. Hall, "The Relation Between Intelligence and Religiosity: A Meta-Analysis and Some Proposed Explanations," *Personality and Social Psychology Review*, October 9, 2013.

Phil Zuckerman cites several studies in his January 2015 op-ed for the *Los Angeles Times* titled "How secular family values stack up": http://www.latimes.com/opinion/op-ed/la-oe-0115-zuckerman-secular-parenting-20150115-story.html.

000 **research has shown that humans will readily sacrifice their lives if they believe their death has meaning.** You can learn more about why people heroically self-sacrifice in Elizabeth

Svoboda, *What Makes a Hero?: The Surprising Science of Selflessness* (New York: Current, 2013).

CHAPTER 5: LOVE SEAT CONFESSIONAL

000 **Brain scientists have found that the further you shift your focus away from yourself and toward an expansive view, the more likely you are to feel this kind of awe.** The link to nature is discussed by Gretchen Reynolds in "How Walking in Nature Changes the Brain," *New York Times*, July 22, 2015: http://well.blogs.nytimes.com/2015/07/22/how-nature-changes-the-brain.

And in a study by Melanie Rudd of Stanford University, Kathleen D. Vohs of the University of Minnesota, and Jennifer Aaker of Stanford University, "Awe Expands People's Perception of Time, Alters Decision Making, and Enhances Well-Being," 2012.

CHAPTER 6: NASA AND BACON NUMBERS

000 **Dryden Flight Research Center** This facility has since been renamed NASA Armstrong Flight Research Center.

000 **There's a study that says 42 percent of Americans will undergo a faith transition at some point in their lives.** Found in Barry A. Kosmin and Ariela Keysar, "American Nones: The Profile of the No Religion Population."

000 **just go watch the trailer for *Blue Like Jazz*, the movie.** I appear on-screen starting at 00:17 in a gray sweater vest and blazer. https://www.youtube.com/watch?v=GOglQgyxYkI.

CHAPTER 7: THE HORSE LEAVES THE BARN

000 **Anthropologist Tanya Luhrmann suggests that American Evangelicals train their brain's reality-sensing mechanism to project some of their own internal life onto an external source.** Her book, *When God Talks Back: Understanding the American Evangelical Relationship with God* (New York: Alfred A. Knopf,

2012) is one of my favorites, and in it she discusses this phenomenon in great detail.

000 **Researchers have found that people who build such elaborate neural networks around God use that network as part of the way they experience reality.** As cited and analyzed by Dr. Andrew Newberg in *How God Changes Your Brain*.

"Neurosurgeons decided to try a radical procedure to treat these patients: they surgically severed the patient's corpus callosum." There's a great overview of this research by David Wolman called "The split brain: A tale of two halves" in *Nature*, March 14, 2012: http://www.nature.com/news/the-split-brain-a-tale-of-two-halves-1.10213.

Neurologist V. S. Ramachandran discusses this research—and the story of a theist and atheist in the same skull—on YouTube. https://youtu.be/PFJPtVRlI64.

CHAPTER 9: EINSTEIN'S GOD

000 **When we look deeply into the night sky with telescopes, we find that all galaxies are moving away from one another and that the rate of their outward movement is increasing.** I play it safe when discussing physics and cosmology, so I'm not going to cite every claim, but I'm well within accepted science in the rest of the book. If you'd like to learn more about how scientists determine the age of the universe, its structure, the nature of the Big Bang or the Singularity, check out these excellent books by actual physicists and cosmologists:

The Fabric of the Cosmos by Brian Greene

A Universe from Nothing by Lawrence Krauss

A Brief History of Time by Stephen Hawking

If those books make your head spin, there's an excellent series of easy-to-follow videos on YouTube called "MinutePhysics."

CHAPTER 10: THE GOD WE CAN KNOW

000 **research coined a term for this discipline: neurotheology.** If you'd like to learn more about how scientists are probing our brains to learn how we relate to God, I'd suggest the following books:

Principles of Neurotheology by Dr. Andrew B. Newberg

How God Changes Your Brain by Dr. Andrew B. Newberg

The God-Shaped Brain by Timothy R. Jennings, M.D.

The "God" Part of the Brain: A Scientific Interpretation of Human Spirituality and God by Matthew Alper.

CHAPTER 11: TEACH US, NEUROSCIENCE, TO PRAY

000 **In one NBC poll** Available in the article "Does prayer work? Is there an afterlife? *TODAY*'s survey offers snapshot of faith, spirituality" by Scott Stump for TODAY.com http://www.today.com/news/there-afterlife-does-prayer-work-todays-survey-faith-spirituality-t14176.

000 **A LifeWay poll found that among people who pray** From "New Research: Americans pray for friends, family, but rarely for celebrities or sports teams" by LifeWay Research http://blog.lifeway.com/newsroom/2014/10/01/new-research-americans-pray-for-friends-family-but-rarely-for-celebrities-or-sports-teams.

000 **Studies have shown that people who pray to God about problems in this way achieve a positive emotional effect, similar to if they'd seen a therapist.** Drawn from Chapter 4 of *When God Talks Back* by Tanya Luhrmann.

000 **research says is even more beneficial: contemplative meditation** Most of the benefits-from-prayer references in this chapter come from *How God Changes Your Brain*, by Andrew Newberg.

CHAPTER 12: JESUS

000 **I was in that studio to have a talk with comedian Pete Holmes for his podcast** *You Made It Weird with Pete Holmes.* Specifically, Episode 201: Science Mike.

000 **I spent a lot of time reading the arguments of the mythicists** Wikipedia's overview on this topic is excellent. https://en.wikipedia.org/wiki/Christ_myth_theory.

CHAPTER 13: TAKE ME TO CHURCH

000 **They show elevated risk for suicide, and some even suffer with Post Traumatic Stress Disorder.** Learn more in "The Health Effects of Leaving Religion" by Jon Fortenbury in *The Atlantic,* September 28, 2014.

000 **But therapy works because human memories are impressionable.** This is well-understood stuff in science today, but you can get an overview by reading "You Memory is like the Telephone Game" by Maria Paul for Northwestern University. http://www.northwestern.edu/newscenter/stories/2012/09/your-memory-is-like-the-telephone-game.html.

000 **When people attempt to shortcut or disavow the sorrow of emotional wounds instead of expressing it . . . they might cause themselves real psychological harm. Experiments show that your intellectual, emotional, and even physical performance can be affected . . .** I discuss this phenomenon in more detail, and cite research, in an article for *RELEVANT* magazine titled "The Scientific Case for Forgiveness." http://www.relevantmagazine.com/life/scientific-case-forgiveness.

000 **Our invention of thrown weapons is mirrored by a sudden collapse of apex-predator populations in every region of Earth populated by humans at the time.** Learn more from *Huffpost Science* in "Early Humans Doomed Large Carnivores Two Million Years Ago, Scientists Say" by Kate Wong. http://www.huffingtonpost.com/2012/04/25/early-humans-large-carnivore_n_1453780.html.

Also, check out "Early humans linked to large-carnivore extinctions" by Jeff Tollefson in *Nature.* http://www.nature.com/news/early-humans-linked-to-large-carnivore-extinctions-1.10508.

000 **Call yourself a liberal, and you'll probably form opinions that largely align with your understanding of what liberals believe.** The power of labels to shape our beliefs is remarkable. Fiske ST. "Bias against outgroups," in Miller, *The Social Psychology of Good and Evil.* Guilford Press, 2004.

CHAPTER 14: THE GOOD BOOK

000 **After all, the Bible isn't a single book, with one voice, one perspective, or one unified take on the history of how God has interacted with His people.** The best resource I've seen for viewing the Bible in an accurate historical posture while still revering it as a sacred work is *The Bible Tells Me So: Why Defending Scripture Has Made Us Unable to Read It* by Peter Enns.

Rob Bell's series "What Is the Bible" is excellent, as well: http://robbellcom.tumblr.com/post/66107373947/what-is-the-bible.

000 **What you may not know is that van Gogh wanted to preach the Gospel and actually sought to become a Methodist minister.** My favorite van Gogh biography is *Van Gogh: The Life* by Steven Naifeh and Gregory White Smith. This pairs well with *Van Gogh's Letters: The Mind of the Artist in Paintings, Drawings, and Words, 1875–1890,* a collection of van Gogh's actual letters, curated by H. Anna Suh.

ABOUT THE AUTHOR

Mike McHargue (better known as Science Mike) is an author, podcaster, and speaker who focuses on the science behind spiritual experiences.

Mike cofounded The Liturgists art and spiritual collective with Michael Gungor of the award-winning band Gungor. Together they assembled a diverse cast of contributors including Rob Bell, Shauna Niequist, Amena Brown, Rachel Held Evans, Pete Holmes, Sleeping At Last, and All Sons and Daughters. Part of The Liturgists' work is a curiously popular podcast, *The Liturgists Podcast.* By examining topics through the lenses of science, art, and faith with a commitment to open, honest discussions, *The Liturgists Podcast* has attracted hundreds of thousands of listeners of remarkable diversity—a broad swath of Christianity, as well as the religiously unaffiliated, including agnostics and atheists.

Mike also hosts a weekly question-and-answer podcast called *Ask Science Mike. Ask Science Mike* is a safe space for people to share questions they've always been afraid to ask about science, faith, and life. Mike also blogs on his website, in addition to writing for *RELEVANT* magazine, Don Miller's Storyline blog, and BioLogos.org.

Mike loves sharing ideas in person and is an in-demand speaker at conferences, churches, and colleges. He's recently spoken to sold-out crowds in New York about the science of addiction, explained the science of storytelling at Storyline in Chicago, and shown people how Christianity can embrace the modern world at Belong in London.

To learn more about Mike's work integrating science and faith, please visit www.mikemchargue.com.

To connect with his work with The Liturgists, please visit www.theliturgists.com.

To bring Mike to talk to your group about how science and faith inform our world, please visit www.mikemchargue.com/speaking.